Set Dynamic Equations on Time Scales

A Brief Introduction with Applications

TRENDS IN ABSTRACT AND APPLIED ANALYSIS

ISSN: 2424-8746

Series Editor: John R. Graef
The University of Tennessee at Chattanooga, USA

This series will provide state of the art results and applications on current topics in the broad area of Mathematical Analysis. Of a more focused nature than what is usually found in standard textbooks, these volumes will provide researchers and graduate students a path to the research frontiers in an easily accessible manner. In addition to being useful for individual study, they will also be appropriate for use in graduate and advanced undergraduate courses and research seminars. The volumes in this series will not only be of interest to mathematicians but also to scientists in other areas. For more information, please go to http://www.worldscientific.com/series/taaa

Published

Trends in Abstract
and Applied Analysis
Volume **13**

Set Dynamic Equations on Time Scales

A Brief Introduction with Applications

Shihuang Hong

Hangzhou Dianzi University, China
Kaili University, China

Sanket Tikare

Ramniranjan Jhunjhunwala College, India

World Scientific

NEW JERSEY · LONDON · SINGAPORE · GENEVA · BEIJING · SHANGHAI · TAIPEI · CHENNAI

Published by

World Scientific Publishing Co. Pte. Ltd.
5 Toh Tuck Link, Singapore 596224
USA office: 27 Warren Street, Suite 401-402, Hackensack, NJ 07601
UK office: 57 Shelton Street, Covent Garden, London WC2H 9HE

Library of Congress Control Number: 2024054714

British Library Cataloguing-in-Publication Data
A catalogue record for this book is available from the British Library.

Trends in Abstract and Applied Analysis — Vol. 13
SET DYNAMIC EQUATIONS ON TIME SCALES
A Brief Introduction with Applications

ISBN 978-981-12-9640-6 (hardcover)
ISBN 978-981-12-9641-3 (ebook for institutions)
ISBN 978-981-12-9642-0 (ebook for individuals)

For any available supplementary material, please visit
https://www.worldscientific.com/worldscibooks/10.1142/13937#t=suppl

Desk Editors: Nambirajan Karuppiah/Lai Fun Kwong

Typeset by Stallion Press
Email: enquiries@stallionpress.com

Preface

The investigation of the theory of set dynamic equations (SDEs), that includes set differential equations and set difference equations, as an independent discipline, has certain advantages (Lakshmikantham *et al.*, 2006). The qualitative and analytical theories of SDEs are developed in the wake of two trends: one is the internal development of the theory of differential equations and other is the attempts to describe real world processes and phenomena giving rise to SDEs. By now there is still no methodology developed for the description of real world phenomena by means of SDEs at the level of engineering applications. Moreover, the general theory of such equations has recently advanced essentially and been growing very rapidly. Nonetheless, the study is still in the initial stages, see for more detail (Martynyuk, 2019).

Like single-valued equations, many results concerning set differential equations carry over quite easily to corresponding results for set difference equations, while other results seem to be completely different in nature from their continuous counterparts. The purpose of this book is to provide a qualitative results for SDEs when the domain of unknown functions is an arbitrary time scale, which allows one to get some insight into and better understanding of the subtle difference between discrete and continuous systems. This proposed study of SDEs avoids proving results twice, once for set differential equations and once for set difference equations.

In this present book, in the framework of general theory of time scales, we present some developed approaches that enable one to study the qualitative theory of set differential and set difference equations in an unified way. The book is organized as follows: Chapter 1 contains some preliminaries, definitions, notations, and concepts concerning time scale calculus. The results in this chapter will cover delta derivative and integrals, and regressive matrices. Chapter 2 assembles the preliminary material providing the necessary tools including the calculus for set-valued maps relevant to the later development. Moreover, we review the general properties of the calculus and inquire into its special properties on the time scale. Chapter 3 is devoted to the investigation of the existence and uniqueness of solutions and the fundamental theory of various SDEs such as impulsive problem, initial value problem, phase space, and almost periodicity. In Chapter 4, the notions of stability for the solutions of SDEs on time scales, using Lyapunov-like functions, are considered. Criteria for the equistability, equiasymptotic stability, uniform, and uniform asymptotic stability are developed. We also present a new definition for certain forms of exponential stability of solutions, including H-exponential stability, H-exponentially asymptotic stability, H-uniformly exponential stability, and H-uniformly exponentially asymptotic stability for a class of SDEs on time scales. Employing Lyapunov-like functions on time scales, we provide sufficient conditions for the exponential stability of the trivial solution for such SDEs. Finally, in Chapter 5 as an application of our results, a class of linear impulsive fuzzy dynamic equations on time scales is studied by using the generalized differentiability concept on time scales. Various novel criteria and general forms of solutions are established for such models whose significance lies in proposing the possibility to get unifying form of solutions for discrete and continuous dynamical systems under uncertainty and to unify corresponding problems in the framework of fuzzy dynamic equations on time scales.

This book is designed for the experts working in the field of time scales theory. However, it can be used as a textbook at the graduate level and as a reference book of several disciplines.

Acknowledgments

Shihuang Hong would like to acknowledge his co-author, *Sanket Tikare*, for his idea for the present book and all his helpful efforts in proofreading an early draft. Shihuang appreciate the knowledge enhancing discussions with him. Shihuang also like to thank his students, *Keke Jia* and *Jieqing Yue*, for assisting him during the writing of the first draft of this monograph. Shihuang extends thanks to his wife *Yingzi Peng* for her support in the writing process of the monograph.

Sanket Tikare acknowledges the valuable and continued support received from *Dr. Usha Mukundan*, Academic Director of Hindi Vidya Prachar Samiti, and *Dr. Himashu Dawda*, Principal of Ramniranjan Jhunjhunwala College, during the preparation of this book. Further, Sanket extends thanks to his wife *Shaila* and daughter *Manasvini* for their moral support and unflinching encouragement. Sanket has dedicated this book to his lifetime teacher *Late Shri Manohar Hari Risbud*.

Both authors wish to express their gratitude to their families and friends for their constant support and encouragement. Finally, authors would like to thank the Editor-in-Chief of the monograph series: Trends in Abstract and Applied Analysis for the World Scientific Publishers, *John R. Graef*, for his interest and continued support to this project, and Executive Editor, *Rochelle Kronzek Miller* and Lai Fun Kwong for their genial attitude and co-operations at every stage in the preparation of the final version of this manuscript.

About the Authors

Shihuang Hong is a Professor of Mathematics in the Department of Mathematics at School of Science, Hangzhou Dianzi University, Hangzhou, P. R. China, and School of Science, Kaili University, Kaili, Guizhou, P. R. China. His main research interests are fixed point theory, ordinary differential equations, set-valued dynamic equations, and Nash equilibrium theory. He published more than 100 academic research papers and supervised more than 30 students for master/doctoral degree. He received several research projects from National Natural Science Foundation of China and Provincial Natural Foundation of China. He won Hainan Province Science and Technology Award.

Sanket Tikare is an Assistant Professor in the Department of Mathematics at Ramniranjan Jhunjhunwala College, which is an empowered autonomous college affiliated with University of Mumbai, Mumbai, Maharashtra, India. He did his MSc and PhD in mathematics both from Shivaji University, Kolhapur, Maharashtra, India in the years 2008 and 2012, respectively. He specializes in Applied Analysis. His research area of interest includes differential equations, difference equations, quantum difference equations, dynamic equations on time scales, impulsive systems, and equations with delays.

Contents

Chapter 1

Essentials of Time Scales Calculus

This chapter is fairly compelling introduction to the basic notions and notations of *time scales calculus*.

1.1 Basic Definitions and Notations

Let $\mathbb{R}, \mathbb{R}^+, \mathbb{Z}$, and $\mathbb{N}$ denote respectively the set of real numbers, non-negative real numbers, integers, and natural numbers. A nonempty closed subset $\mathbb{T}$ of $\mathbb{R}$ is called a time scale or a measure chain. An excellent introduction and description of calculus on time scales is given in Bohner and Peterson (2001, 2003). The contemporary advances in the field of dynamic equations on time scales and their applications can be found in Georgiev (2023) and Agarwal *et al.* (2024). In what follows, by $I_{\mathbb{T}}$ we mean the set $I \cap \mathbb{T}$ for $I \subset \mathbb{R}$.

Let $\mathbb{T}$ be a time scale. For $t \in \mathbb{T}$, we define the forward jump operator $\sigma \colon \mathbb{T} \to \mathbb{T}$ by

$$\sigma(t) = \inf\{\tau \in \mathbb{T} \colon \tau > t\},$$

while the backward jump operator $\rho \colon \mathbb{T} \to \mathbb{T}$ is defined by

$$\rho(r) = \sup\{\tau \in \mathbb{T} \colon \tau < r\}.$$

In this definition, we put $\inf \emptyset = \sup \mathbb{T}$ (i.e., $\sigma(M) = M$ if $\mathbb{T}$ has a maximum M) and $\sup \emptyset = \inf \mathbb{T}$ (i.e., $\rho(m) = m$ if $\mathbb{T}$ has a minimum m), where $\emptyset$ denotes the empty set.

A point t is said to be

- right-scattered if $\sigma(t) > t$;
- right-dense (rd) if $t < \sup \mathbb{T}$ and $\sigma(t) = t$;

- left-scattered if $\rho(t) < t$;
- left-dense (ld) if $t > \inf \mathbb{T}$ and $\rho(t) = t$;
- isolated if it is right-scattered and left-scattered at the same time;
- dense if it is right-dense and left-sense at the same time.

We introduce two sets $\mathbb{T}^\kappa$ and $\mathbb{T}_\kappa$ which are derived from the time scale $\mathbb{T}$ as follows. If $\mathbb{T}$ has a right-scattered minimum m, then $\mathbb{T}_\kappa = \mathbb{T} \setminus \{m\}$; otherwise set $\mathbb{T}_k = \mathbb{T}$. If $\mathbb{T}$ has a left-scattered maximum M, then $\mathbb{T}^\kappa = \mathbb{T} \setminus \{M\}$; otherwise, set $\mathbb{T}^\kappa = \mathbb{T}$. In this book, for the sake of the convenience, we always write $\mathbb{T}$ instead of $\mathbb{T}^\kappa$ or $\mathbb{T}_\kappa$, unless specifically stated. The graininess function $\mu \colon \mathbb{T} \to \mathbb{R}^+$ is defined by

$$\mu(t) = \sigma(t) - t.$$

A function $f \colon \mathbb{T} \to \mathbb{R}$ is right-dense continuous (rd-continuous, for short) provided f is continuous at each right-dense point in $\mathbb{T}$ and its left-sided limits exist at each left-dense points in $\mathbb{T}$. The set of all right-dense continuous functions from $\mathbb{T}$ to $\mathbb{R}$ is denoted by $\mathcal{C}_{\mathrm{rd}}(\mathbb{T}, \mathbb{R})$ or $\mathcal{C}_{\mathrm{rd}}$. Similarly, we can define left-dense continuous (ld-continuous) function. The set of all left-dense continuous functions from $\mathbb{T}$ to $\mathbb{R}$ is denoted by $\mathcal{C}_{\mathrm{ld}}(\mathbb{T}, \mathbb{R})$ or $\mathcal{C}_{\mathrm{ld}}$.

A function $f \colon \mathbb{T} \to \mathbb{R}$ is called regulated provided its right-sided limits exist (belong to $\mathbb{R}$) at all right-dense points in $\mathbb{T}$ and its left-sided limits exist (belong to $\mathbb{R}$) at all left-dense points in $\mathbb{T}$.

A function $f \colon \mathbb{T} \to \mathbb{R}$ is called regressive provided

$$1 + \mu(t)f(t) \neq 0$$

for all $t \in \mathbb{T}^\kappa$. A function $f \colon \mathbb{T} \to \mathbb{R}$ is called positively regressive provided

$$1 + \mu(t)f(t) > 0$$

for all $t \in \mathbb{T}^\kappa$. The following notation will be used:

$$\mathcal{R} = \{p \colon \mathbb{T} \to \mathbb{R} \colon p \text{ is an rd-continuous regressive function}\}.$$

and

$$\mathcal{R}^+ = \{p \colon \mathbb{T} \to \mathbb{R} \colon p \text{ is an rd-continuous and positively regressive function}\}.$$

Furthermore, for $p, q \in \mathcal{R}$, the "circle plus" and "circle minus" are defined as, respectively,

$$p \oplus q = p + q + \mu pq,$$

$$p \ominus q = \frac{p - q}{1 + \mu q}, \quad \text{and}$$

$$\ominus p = 0 \ominus p.$$

1.2 Delta Derivatives and Integrals

The delta derivative of function $f \colon \mathbb{T} \to \mathbb{R}$ at point $t \in \mathbb{T}$ is defined as follows.

Definition 1.1. Let $f \colon \mathbb{T} \to \mathbb{R}$ be a given function and $t \in \mathbb{T}^{\kappa}$. The number $f^{\Delta}(t)$ (whenever it exists) is called the delta (or Hilger) derivative of f at t provided for each $\varepsilon > 0$, there exists a neighborhood $U_{\mathbb{T}}$ of t (i.e., $U_{\mathbb{T}} = (t - \delta, t + \delta) \cap \mathbb{T}$ for some $\delta > 0$) such that

$$|f(\sigma(t)) - f(s) - f^{\Delta}(t)(\sigma(t) - s)| \leq \varepsilon |\sigma(t) - s|,$$

for all $s \in U_{\mathbb{T}}$.

We say f is delta differentiable (or Δ-differentiable) at t provided its delta derivative exists at t. Moreover, we say f is delta differentiable on $\mathbb{T}^{\kappa}$ provided its delta derivative exists at each $t \in \mathbb{T}^{\kappa}$. The function $f^{\Delta} \colon \mathbb{T}^{\kappa} \to \mathbb{R}$ is called the delta (or Hilger) derivative of f on $\mathbb{T}^{\kappa}$.

Remark 1.1. For $\mathbb{T} = \mathbb{R}$, $f^{\Delta}(t) = f'(t)$, the ordinary derivative, and for $\mathbb{T} = \mathbb{Z}$, $f^{\Delta}(t) = f(t + 1) - f(t)$ which is the forward difference.

Definition 1.2. A continuous function $f \colon \mathbb{T} \to \mathbb{R}$ is called pre-differentiable with (region of differentiation) D, provided

(1) $D \subset \mathbb{T}^{\kappa}$;
(2) $\mathbb{T}^{\kappa} \setminus D$ is countable and contains no right-scattered elements of $\mathbb{T}$;
(3) f is differentiable on D.

Lemma 1.1. *Suppose $f_n \colon \mathbb{T} \to \mathbb{R}$ is pre-differentiable with $D \subset \mathbb{T}$ for each $n \in \mathbb{N}$. Assume that for each $t \in \mathbb{T}$ there exists a compact* *interval neighborhood $U(t)$ such that the sequence $\{f_n^\Delta\}_{n\in\mathbb{N}}$ converges uniformly on $U(t) \cap D$. Then,*

(i) *$\{f_n\}$ converges uniformly on $U(t)$ provided $\{f_n\}$ converges at some $t_0 \in U(t)$ for some $t \in \mathbb{T}$.*

(ii) *$\{f_n\}$ converges uniformly on $U(t)$ for some $t \in \mathbb{T}$ provided $\{f_n\}$ converges at some $t_0 \in \mathbb{T}$.*

(iii) *the limit function $f = \lim_{n\to\infty} f_n$ is pre-differentiable with D, and we have*

$$f^\Delta(t) = \lim_{n\to\infty} f_n^\Delta(t)$$

for all $t \in D$.

Theorem 1.1 (Existence of pre-antiderivatives). *Let f be regulated. Then there exists a function F which is pre-differentiable with region of differentiation D such that*

$$F^\Delta(t) = f(t) \quad \text{holds for all} \quad t \in D.$$

Now, if a single-valued function F is delta differentiable on an interval $D \subset \mathbb{T}$ and

$$F^\Delta(t) = f(t) \quad \text{for} \quad t \in D,$$

we define the Cauchy integral by

$$\int_a^t f(s)\Delta s = F(t) - F(a).$$

In this case, we say f to be delta integrable on D. In particular, by

$$\int_a^\infty f(t)\Delta t := \lim_{b\to\infty} \int_a^b f(t)\Delta t$$

we mean that f is delta integrable on $D = [a, \infty)_\mathbb{T}$ provided that this limit exists.

Theorem 1.2 (Mean value theorem). *Let f and g be real-valued functions defined on $\mathbb{T}$, both pre-differentiable with D. Then,*

$$|f^\Delta(t)| \le g^\Delta(t) \quad \text{for all} \quad t \in D$$

implies

$$|f(s) - f(r)| \le g(s) - g(r) \quad \text{for all} \quad r, s \in \mathbb{T}, \quad r \le s.$$

Corollary 1.1. *Suppose f and g are pre-differentiable with D. If U is a compact interval with endpoints $r, s \in \mathbb{T}$, then*

$$|f(s) - f(r)| \leq \left\{ \sup_{t \in U^\kappa \cap D} |f^\Delta(t)| \right\} |s - r|.$$

Theorem 1.3. *Assume that $\nu \colon \mathbb{T} \to \mathbb{R}$ is strictly increasing and $\widetilde{\mathbb{T}} := \nu(\mathbb{T})$ is a time scale. Let $w \colon \widetilde{\mathbb{T}} \to \mathbb{R}$. If $\nu^\Delta(t)$ and $w^{\widetilde{\Delta}}(\nu(t))$ exist for $t \in \mathbb{T}^\kappa$, then*

$$(w \circ \nu)^\Delta = (w^{\widetilde{\Delta}} \circ \nu)\nu^\Delta.$$

Theorem 1.4 (Derivative of the inverse). *Assume $\nu \colon \mathbb{T} \to \mathbb{R}$ is strictly increasing and $\widetilde{\mathbb{T}} := \nu(\mathbb{T})$ is a time scale. Then,*

$$\frac{1}{\nu^\Delta} = (\nu^{-1})^{\widetilde{\Delta}} \circ \nu$$

at points where ν^Δ is different from zero.

Theorem 1.5. *If $a, b, c \in \mathbb{T}$, $\alpha \in \mathbb{R}$, and $f, g \in \mathcal{C}_{\mathrm{rd}}$, then*

(i)
$$\int_a^b [f(t) + g(t)]\Delta t = \int_a^b f(t)\Delta t + \int_a^b g(t)\Delta t;$$

(ii)
$$\int_a^b (\alpha f(t))\Delta t = \alpha \int_a^b f(t)\Delta t;$$

(iii)
$$\int_a^b f(t)\Delta t = -\int_b^a f(t)\Delta t;$$

(iv)
$$\int_a^b f(t)\Delta t = \int_a^c f(t)\Delta t + \int_c^b f(t)\Delta t;$$

(v)
$$\int_a^b f(\sigma(t))g^\Delta(t)\Delta t = (fg)(b) - (fg)(a) - \int_a^b f^\Delta(t)g(t)\Delta t;$$

(vi)

$$\int_a^b f^\Delta(t)g(t)\Delta t = (fg)(b) - (fg)(a) - \int_a^b f(\sigma(t))g^\Delta(t)\Delta t;$$

(vii)

$$\int_a^a f(t)\Delta t = 0;$$

(viii) *if* $|f(t)| \leq g(t)$ *on* $[a, b)_{\mathbb{T}}$, *then*

$$\left| \int_a^a f(t)\Delta t \right| \leq \int_a^b g(t)\Delta t;$$

(ix) *if* $f(t) \geq 0$ *for all* $a \leq t < b$, *then*

$$\int_a^a f(t)\Delta t \geq 0.$$

Definition 1.3. For $p \in \mathcal{R}$, we define the exponential function by

$$e_p(t, s) = \exp\left(\int_s^t \xi_{\mu(t)}(p(\tau))\Delta\tau \right) \quad \text{for } s, t \in \mathbb{T},$$

where

$$\xi_h(z) \colon \left\{ z \in \mathbb{C} \colon z \neq \frac{1}{h} \right\} \to \left\{ z \in \mathbb{C} \colon \frac{-\pi}{h} < \operatorname{Im}(z) \leq \frac{\pi}{h} \right\}, \quad h > 0,$$

is the cylinder transformation given by

$$\xi_h(z) = \begin{cases} \frac{\log(1+zh)}{h} & \text{for } h \neq 0 \\ z & \text{for } h = 0. \end{cases}$$

In Bohner and Peterson (2001), it is shown that $e_p(\cdot, s)$ with $s \in \mathbb{T}$, is the unique solution to the dynamic initial value problem

$$y^\Delta = p(t)y, \quad y(s) = 1.$$

Theorem 1.6. *If* $p, q \in \mathcal{R}$, *then*

(i) $e_0(t, s) \equiv 1$ *and* $e_p(t, t) \equiv 1$;

(ii) $e_p(\sigma(t), s) = (1 + \mu(t)p(t))e_p(t, s)$;

(iii) $\frac{1}{e_p(t,s)} = e_{\ominus p}(t, s)$;

(iv) $e_p(t, s) = \frac{1}{e_p(s,t)}$;

(v) $e_p(t,s)e_p(s,r) = e_p(t,r)$;

(vi) $e_p(t,s)e_q(t,s) = e_{p\oplus q}(t,s)$;

(vii) $\frac{e_p(t,s)}{e_q(t,s)} = e_{p\ominus q}(t,s)$;

(viii) $\left(\frac{1}{e_p(\cdot,s)}\right)^\Delta = -\frac{p(t)}{e_p^\sigma(\cdot,s)}$.

Theorem 1.7 (Sign of the exponential function). *Let $p \in \mathcal{R}$ and $t_0 \in \mathbb{T}$.*

(i) *If $p \in \mathcal{R}^+$, then $e_p(t,t_0) > 0$ for all $t \in \mathbb{T}$.*

(ii) *If $1 + \mu(t)p(t) < 0$ for some $t \in \mathbb{T}^\kappa$, then*

$$e_p(t,t_0)e_p(\sigma(t),t_0) < 0.$$

(iii) *If $1 + \mu(t)p(t) < 0$ for all $t \in \mathbb{T}^\kappa$, then $e_p(t,t_0)$ changes sign at every point $t \in \mathbb{T}$.*

1.3   Regressive Matrices

We recall some useful concepts regarding matrices. An $m \times n$-matrix-valued function A is said to be delta differentiable on $\mathbb{T}$ provided each entry of A is delta differentiable on $\mathbb{T}$. In this case, we put

$$A^\Delta = (a_{ij}^\Delta)_{1 \leq i \leq m,\, i \leq j \leq n}, \quad \text{where } A = (a_{ij}).$$

Theorem 1.8. *If A is delta differentiable at $t \in \mathbb{T}^\kappa$, then*

$$A^\sigma(t) = A(t) + \mu(t)A^\Delta(t).$$

An $n \times n$-matrix-valued function A on $\mathbb{T}$ is called regressive provided

$$\mathcal{I} + \mu(t)A(t)$$

is invertible for all $t \in \mathbb{T}$. Here, $\mathcal{I}$ stands for an $n \times n$-identity matrix. Let $\mathcal{R}_n = \mathcal{R}_n(\mathbb{T}, \mathbb{R}^{n \times n})$ denotes the set of all regressive rd-continuous $n \times n$-matrix-valued functions from $\mathbb{T}$ to $\mathbb{R}^{n \times n}$ and $\mathcal{R}_n^+$ denotes the set of all positively regressive rd-continuous $n \times n$-matrix-valued functions from $\mathbb{T}$ to $\mathbb{R}^{n \times n}$. From now on, unless otherwise mentioned, the matrix-valued functions involved in equations are always assumed to belong to $\mathcal{R}_n$.

For $A, B \in \mathcal{R}_n$, the "circle plus" and "circle minus" of matrix-valued functions are referred to as, respectively,

$$(A \oplus B)(t) = A(t) + B(t) + \mu(t)A(t)B(t),$$
$$(A \ominus B)(t) = (A \oplus (\ominus B))(t),$$

where

$$(\ominus A)(t) = -[\mathcal{I} + \mu(t)A(t)]^{-1}A(t)$$
$$= -A(t)[\mathcal{I} + \mu(t)A(t)]^{-1}.$$

A matrix exponential function $e_A(t, t_0)$ is defined as the unique matrix-valued solution of the dynamic initial value problem

$$Y^{\Delta} = A(t)Y, \quad Y(t_0) = \mathcal{I},$$

where $A \in \mathcal{R}_n$ is an $n \times n$-matrix-valued function and $t_0 \in \mathbb{T}$. Denote

$$\mathscr{A} = \{e_B(u, v) \colon B \colon \mathbb{T} \to \mathbb{R}^{n \times n} \text{ is an } n \times n\text{-matrix-valued function}$$

$$\text{and } u, v \in \mathbb{T}\}.$$

Remark 1.2. Assume that A is a constant $n \times n$ matrix. If $\mathbb{T} = \mathbb{R}$, then

$$e_A(u, v) = e^{A(u-v)},$$

while if $\mathbb{T} = \mathbb{Z}$ and $\mathcal{I} + A$ is invertible, then

$$e_A(u, v) = (\mathcal{I} + A)^{u-v}$$

and $e_A(u, v) \in \mathscr{A}$.

In Bohner and Peterson (2001), the elements of $\mathscr{A}$ have been proved to possess the properties listed in the following theorem.

Theorem 1.9. *If $e_A(\cdot, \cdot), e_B(\cdot, \cdot) \in \mathscr{A}$, then*

(i) $e_0(t, s) \equiv \mathcal{I}, \ e_A(t, t) \equiv \mathcal{I}, \ and \ e_A(\sigma(t), s) = (\mathcal{I} + \mu(t)A(t))e_A(t, s);$

(ii) $e_A(s, t) = e_A^{-1}(t, s) = e_{\ominus A^*}^*(t, s),$ where A^* stands for the conjugate transpose of the matrix A;

(iii) $e_A(t, s)e_A(s, r) = e_A(t, r);$

(iv) $e_{A \oplus B}(t, s) = e_A(t, s)e_B(t, s)$ provided $e_A(t, s)$ and $B(t)$ commute;

(v) $e_A^\Delta(c, \cdot) = -e_A(c, \sigma(\cdot))A$ *for* $c \in \mathbb{T}$;

(vi) $\int_a^b e_A(c, \sigma(t))A(t)\Delta t = e_A(c, a) - e_A(c, b)$ *for* $a, b, c \in \mathbb{T}$.

Theorem 1.10 (Variation of constants). *Let* $A \in \mathcal{R}$ *be an* $n \times n$-*matrix-valued function on* $\mathbb{T}$ *and suppose that* $f \colon \mathbb{T} \to \mathbb{R}^n$ *is rd-continuous. Let* $t_0 \in \mathbb{T}$ *and* $y_0 \in \mathbb{R}^n$. *Then the dynamic initial value problem (DIVP)*

$$\begin{cases} y^\Delta = A(t)y + f(t), & t \in \mathbb{T}, \\ y(t_0) = y_0 \end{cases}$$

has a unique solution $y \colon \mathbb{T} \to \mathbb{R}^n$. *Moreover, this solution is given by*

$$y(t) = e_A(t, t_0)y_0 + \int_{t_0}^t e_A(t, \sigma(\tau))f(\tau)\Delta\tau.$$

Theorem 1.11 (Putzer algorithm). *Let* $A \in \mathcal{R}$ *be a constant* $n \times n$-*matrix-valued function. Suppose* $t_0 \in \mathbb{T}$. *If* $\lambda_1, \lambda_2, \ldots, \lambda_n$ *are eigenvalues of* A, *then*

$$e_A(t, t_0) = \sum_{i=0}^{n-1} r_{i+1}(t)P_i,$$

where $r(t) := (r_1(t), r_2(t), \ldots, r_n(t))^T$ *is the solution of the DIVP*

$$r^\Delta = \begin{pmatrix} \lambda_1 & 0 & 0 & \cdots & 0 \\ 1 & \lambda_2 & 0 & \ddots & \vdots \\ 0 & 1 & \lambda_3 & \ddots & \vdots \\ \vdots & \ddots & \ddots & \ddots & 0 \\ 0 & \cdots & 0 & 1 & \lambda_n \end{pmatrix} r, \quad r(t_0) = \begin{pmatrix} 1 \\ 0 \\ 0 \\ \vdots \\ 0 \end{pmatrix},$$

and the P-*matrices* $P_0, P_1, \ldots, P_n$ *are recursively defined by* $P_0 = \mathcal{I}$ *and*

$$P_{k+1} = (A - \lambda_{k+1}\mathcal{I})P_k \quad for\ 0 \le k \le n - 1.$$

Theorem 1.12 (Bernoulli's inequality). *Let* $\alpha \in \mathbb{R}$ *with* $\alpha \in \mathcal{R}^+$. *Then,*

$$e_\alpha(t, s) \ge 1 + \alpha(t - s) \quad for\ all\ t \ge s.$$

Theorem 1.13 (Gronwall's inequality). *Let $y, f \in \mathcal{C}_{\mathrm{rd}}(\mathbb{T}, \mathbb{R})$ and $p \in \mathcal{R}^+$, $p \geq 0$. Then,*

$$y(t) \leq f(t) + \int_{t_0}^{t} y(\tau)p(\tau)\Delta\tau \quad \text{for all } t \in \mathbb{T}$$

implies

$$y(t) \leq f(t) + \int_{t_0}^{t} e_p(t, \sigma(\tau))y(\tau)p(\tau)\Delta\tau \quad \text{for all } t \in \mathbb{T}.$$

Chapter 2

Set Functional Calculus

2.1   Introduction

From the technical point of view, similar to dealing with single-valued cases, it is worth avoiding proving results twice, once for set differential equations and once for set difference equations. This inspired us to formulate a class of new time-scale calculus of set-valued functions to exhibit the two main features of unification and extension, as an analogy of single-valued dynamic equations on time scales. Hence, we first extend the Hukuhara derivative for time scales which is of basic importance and plays a fundamental role in both theoretical and applied studies of set dynamic equations on time scales (SDEs). To achieve our purpose, a new derivative of set-valued functions on time scale is introduced to suit our study of set dynamic equations. The proposed approach forms the appropriate environment for the study of set dynamic systems on time scales. Moreover, some corresponding properties are explored, providing the necessary background for further consideration.

In this chapter, we recall the theory of calculus for set-valued functions defined on time scale. We refer to Hong (2009) and Stefanini and Bede (2009) for details. The notions of Hukuhara derivative and geometric difference play a fundamental role in the theory of SDEs. To achieve our purpose, we adopt a generalization of geometric difference proposed in Stefanini and Bede (2009) that aims to guarantee the existence of difference for any two sets and the derivative of set-valued functions on time scale proposed in Hong and Peng (2016), Δ_g-derivative for short, which is an extension of

Δ_H-derivative introduced in Hong (2009). This will be developed to suit to our study of SDEs. The proposed approach forms the appropriate environment within which the theory of SDEs on time scales can be developed. Moreover, some corresponding properties of Δ_g-derivative are explored which provide the necessary background for our further consideration.

2.2   Hausdorff Metrics

We describe the basic known results of Hausdorff metrics, continuity and differentiability for set-valued functions on time scales and their corresponding properties within the framework of time scales. We refer readers to Lakshmikantham *et al.* (2006) for details. Let $\mathbb{R}^n(n \in \mathbb{N})$ denote the n-dimensional real vector space ($\mathbb{R}^1 = \mathbb{R}$, a real number set) and K_c^n denote the collection of nonempty compact and convex subsets of $\mathbb{R}^n$. The following operations can be naturally defined on K_c^n. For $X, Y \in K_c^n$,

$$X + Y := \{x + y \colon x \in X, y \in Y\},$$
$$\lambda X := \{\lambda x \colon x \in X\}, \quad \lambda \in \mathbb{R},$$
$$XY := \{xy \colon x \in X, y \in Y\}.$$

Here, assume that the product operation in $\mathbb{R}^n$ is well defined. The set $Z \in K_c^n$ satisfying $X = Y + Z$ is known as the geometric difference or Hukuhara difference (H-difference) of set X and set Y and is denoted by the symbol $X-_HY$. It is worthy to note that the geometric difference of two sets does not always exist but if it does it is unique.

A generalization of geometric difference proposed in Stefanini and Bede (2009) aims to guarantee the existence of difference for any two intervals in K_c^1. In the light of this, a generalized difference called the g-difference, "$-_g$", can be defined, for any $X, Y \in K_c^n$, as follows:

$$X-_gY = Z \Leftrightarrow \begin{cases} \text{(a)} & X = Y + Z, \quad \text{or} \\ \text{(b)} & Y = X + (-1)Z. \end{cases} \tag{2.1}$$

It is clear if the g-difference exists, it is unique and it is a generalization of the geometric difference. Further, $X-_gY = X-_HY$, whenever $X-_H Y$ exists. In addition, we have the following result enlisting useful properties.

Proposition 2.1. *Let* $X, Y \in K_c^n$. *Then we have the following:*

  (i) $X-_g X = \{0\}$, $(X + Y)-_g Y = X$, *and* $\{0\}-_g(X-_g Y) = (-Y)-_g(-X)$, *where* $-A = (-1)A$ *for* $A \in K_c^n$.
 (ii) *If* $X-_g Y$ *exists in the sense* (a) *of* (2.1), *then* $Y -_g X$ *exists in the sense* (b) *of* (2.1) *and vice versa.*
(iii) $X-_g Y = Y-_g X = Z$ *if and only if* $Z = \{0\}$ *and* $X = Y$.
(iv) $X-_g Y$ *exists if and only if both* $Y-_g X$ *and* $(-Y)-_g(-X)$ *exist. Further,* $-(X-_g Y) = (-1)((-Y) -_g(-X)) = Y-_g X$.

Proof. We give the proof of (iv) only. To prove the first part of (iv), let $Z = X-_g Y$, that is, $X = Y + Z$ or $Y = X + (-1)Z$. Then, $-X = -Y + (-1)Z$ or $-Y = -X - (-1)Z$ and this means $(-Y)-_g(-X) = -Z$, the second part is immediate. $\qquad\square$

Throughout this book, we always assume that the g-difference of any two considered elements in K_c^n exists. We remark that the assumption may be valid, for instance, in the unidimensional case (with $K_c^1 = \mathbb{I}$, a class of all closed bounded intervals of the real line), the g-difference exists for any two compact intervals.

We define the Hausdorff metric as

$$D[X,Y] = \max\left\{\sup_{y\in Y} d(X,y), \sup_{x\in X} d(x,Y)\right\},$$

where X and Y are bounded subsets of $\mathbb{R}^n$ and $d(x, Y) := \inf\{d(x,y) : y \in Y\}$ and $d(X,y) := \inf\{d(x,y) : x \in X\}$. It is obvious that the Hausdorff metric satisfies conditions of the ordinary metric d.

Note that K_c^n with this distance D is a complete metric space. In fact, K_c^n is a semilinear complete metric space. On the other hand, the Hausdorff metric D is compatible with the operations defined on it as described by the following properties. For $X, Y, Z, W \in K_c^n$ and $\mu, \nu \in \mathbb{R}$, we have

  (i) $D[X \pm_g Z, Y \pm_g Z] = D[X, Y]$,
 (ii) $D[X \pm_g Z, Y \pm_g W] \le D[X, Y] + D[Z, W]$,
(iii) $D[\mu X, \mu Y] = |\mu| D[X, Y]$,

(iv) $D[XZ, YZ] \leq \|Z\| D[X, Y]$,

(v) $D[\mu X, \nu X] = |\mu - \nu| \|X\|$ for $\mu, \nu \geq 0$ or $\mu, \nu \leq 0$.

Here $\|V\| := D[V, \theta] = \sup\{\|v\| : v \in V\}$ for $V \in K_c^n$ and, throughout this book, θ stands for the zero element of K_c^n, regarded as the set $\{0\}$ containing one point.

Lemma 2.1. *If $X, Y \in K_c^n$ and $x \in X$, then, for any positive number $q < 1$, there exists $y = y(x) \in Y$ such that*

$$qd(x, y) \leq D[X, Y].$$

Definition 2.1. We can regard a set-valued function F as a correspondence that maps a point in time scale $\mathbb{T}$ to a subset in K_c^n. We can write $F \colon \mathbb{T} \to K_c^n$. The set-valued function is also known as multifunction or multimapping.

In order to define the continuity and regularity of set-valued functions on time scales $\mathbb{T}$, we first need the following notion.

Definition 2.2. Let $\mathbf{D} \subset \mathbb{T}$. A function $f \colon \mathbf{D} \to \mathbb{R}^n$ is called a selector of the set-valued function $F \colon \mathbf{D} \to K_c^n$ provided $f(t) \in F(t)$ for all $t \in \mathbf{D}$.

Definition 2.3. Let $J_\mathbb{T} = J \cap \mathbb{T}$ with $J \subset \mathbb{R}$. A set-valued function $F \colon J_\mathbb{T} \to K_c^n$ is said to have a limit at $t_0 \in J_\mathbb{T}$ provided there exists an element $\mathcal{A} \in K_c^n$ such that, for any given $\varepsilon > 0$, there exists $\delta = \delta(\varepsilon, t_0) > 0$ such that

$$D[F(t), \mathcal{A}] < \varepsilon$$

for all $t \in J_\mathbb{T}$ with $|t - t_0| < \delta$. We denote the limit by

$$\lim_{t \to t_0} F(t) = \mathcal{A}.$$

Let $F(t_0)$ be well defined. Then, F is called continuous at $t_0 \in J_\mathbb{T}$ provided its limit at t_0 exists and equals $F(t_0)$.

A set-valued function $F \colon \mathbb{T} \to K_c^n$ is called regulated provided its right-sided limit exists at right-dense point in $\mathbb{T}$, its left-sided limit exists at left-dense point in $\mathbb{T}$, and its regulated selector exists.

A set-valued function F is called right-dense continuous (rd-continuous), provided F is continuous at each right-dense point in $\mathbb{T}$, its left-sided limits exist at each left-dense point in $\mathbb{T}$, and its rd-continuous selector exists. Similarly, we can define the ld-continuity.

A set-valued function F is said to be uniformly rd-continuous on $\mathbf{D} \subset \mathbb{T}$ provided it is rd-continuous and for any given $\varepsilon > 0$, there exists $\delta > 0$ such that

$$D[F(t), F(s)] < \varepsilon$$

for each right-dense point $t \in \mathbf{D}$ and any $s \in \mathbf{D}$ with $|s - t| < \delta$.

Lemma 2.2. *Let* $F \colon [a,b]_\mathbb{T} \to K_c^n$ *be regulated, where* $[a,b]_\mathbb{T}$ *is a compact interval in* $\mathbb{T}$. *Then,* F *is bounded, i.e., there exists a positive number* M *such that*

$$\|F(t)\| \leq M \quad \text{for each} \quad t \in [a,b]_\mathbb{T}.$$

Proof. Assume that F is unbounded. Then, for each $n \in \mathbb{N}$, there exists $t_n \in [a,b]_\mathbb{T}$ with

$$D[F(t_n), \{0\}] > n.$$

Since $\{t_n \colon n \in \mathbb{N}\} \subset [a,b]_\mathbb{T}$, there exists a convergent subsequence $\{t_{n_k}\}$. Hence, there exists $t_0 \in [a,b]_\mathbb{T}$ such that

$$\lim_{k \to \infty} t_{n_k} = t_0.$$

Since $\mathbb{T}$ is closed and $\{t_{n_k}\} \subset \mathbb{T}$, $t_0 \in \mathbb{T}$. This yields that t_0 cannot be isolated. Therefore, there exists either a subsequence that tends to t_0 from above or a subsequence that tends to t_0 from below. In any case, the limit of $F(t)$ as $t \to t_0$ has to be finite according to regularity, i.e., $D[F(t_0), \{0\}] < \infty$. Thus,

$$D[F(t_n), \{0\}] \leq D[F(t_n), F(t_0)] + D[F(t_0), \{0\}] < \infty$$

for large enough $n \in \mathbb{N}$. This means that F is bounded, which is contradiction. This completes the proof. $\qquad\square$

2.3 Calculus

2.3.1 *Derivative of set-valued functions*

We introduce the Δ_H-derivative for set-valued function under H-difference as follows.

Definition 2.4. Assume that $F \colon \mathbb{T} \to K_c^n$ is a set-valued function and $t \in \mathbb{T}^\kappa$. Let $\Delta_H F(t)$ be an element of K_c^n (provided it exists) with the property that, for given any $\varepsilon > 0$, there exists a neighborhood $U_\mathbb{T}$ of t (i.e., $U_\mathbb{T} = (t - \delta, t + \delta) \cap \mathbb{T}$ for some $\delta > 0$) such that

$$D[F(t + h) -_H F(\sigma(t)), \Delta_H F(t)(h - \mu(t))] \leq \varepsilon |h - \mu(t)|$$

and

$$D[F(\sigma(t)) -_H F(t - h), \Delta_H F(t)(\mu(t) + h)] \leq \varepsilon |\mu(t) + h|$$

for all $t - h, t + h \in U_\mathbb{T}$ with $0 \leq h < \delta$, where $\mu(t) = \sigma(t) - t$. Then, $\Delta_H F(t)$ is said to be the Δ_H-derivative of F at t.

We say that F is Δ_H-differentiable at t, whenever its Δ_H-derivative exists at t. Moreover, we say F is Δ_H-differentiable on $\mathbb{T}$ provided its Δ_H-derivative exists at each $t \in \mathbb{T}^\kappa$. The set-valued function $\Delta_H F \colon \mathbb{T} \to K_c^n$ is then called the Δ_H-derivative of F on $\mathbb{T}$.

Note 2.1. Δ_H-differentiability is also known as delta Hukuhara differentiability.

Proposition 2.2. *If the Δ_H-derivative of F at $t \in \mathbb{T}^\kappa$ exists, then it is unique. Hence, the Δ_H-derivative is well defined.*

Proof. Suppose $F \colon \mathbb{T} \to K_c^n$ is Δ_H-differentiable at $t \in \mathbb{T}^\kappa$. Let $^1\Delta_H F(t)$ and $^2\Delta_H F(t)$ both be the Δ_H-derivative of F at t. Then,

$$D[^1\Delta_H F(t), {}^2\Delta_H F(t)]$$

$$= \frac{1}{|h - \mu(t)|} D[^1\Delta_H F(t)(h - \mu(t)), {}^2\Delta_H F(t)(h - \mu(t))]$$

$$\leq \frac{1}{|h - \mu(t)|} D[^1\Delta_H F(t)(h - \mu(t)), F(t + h) -_H F(\sigma(t))]$$

$$+ \frac{1}{|h - \mu(t)|} D[F(t + h) -_H F(\sigma(t)), {}^2\Delta_H F(t)(h - \mu(t))]$$

$$\leq \varepsilon + \varepsilon.$$

Thus,

$$D[^1\Delta_H F(t), {}^2\Delta_H F(t)] \le 2\varepsilon \quad \text{for} \quad |\mu(t) - h)| \ne 0.$$

From the arbitrariness of $\varepsilon > 0$, it follows that

$$D[^1\Delta_H F(t), {}^2\Delta_H F(t)] = 0.$$

This implies that

$$^1\Delta_H F(t) = {}^2\Delta_H F(t).$$

That is, $^1\Delta_H F = {}^2\Delta_H F$. This completes the proof. $\qquad\square$

Example 2.1. If $F\colon \mathbb{T} \to K_c^1$ is defined by $F(t) = [0, a]$ for all $t \in \mathbb{T}$, where $a \in \mathbb{R}^+$ is constant, then $\Delta_H F(t) = \theta$. Clearly, for any given $\varepsilon > 0$, we have

$$D[F(t + h) -_H F(\sigma(t)), \theta(h - \mu(t))] = D[[0, a] - [0, a], \theta]$$
$$= 0.$$

Hence,

$$D[F(t + h) -_H F(\sigma(t)), \theta(h - \mu(t))] \le \varepsilon |h - \mu(t)| \quad \text{for all} \quad h \ge 0.$$

Similarly, we have $D[F(\sigma(t)) -_H F(t - h), \theta(\mu(t) + h)] = 0 \le \varepsilon |\mu(t) + h|$.

Example 2.2. If $F\colon [a, +\infty)_\mathbb{T} \to K_c^n$ is defined by $F(t) = [a, t]$ for all $t \in [a, +\infty)_\mathbb{T}$, then $\Delta_H F(t) \equiv [0, 1]$. Clearly, for any given $\varepsilon > 0$, we have

$$D[F(t + h) -_H F(\sigma(t)), [0, 1](h - \mu(t))]$$
$$= D[[a, t + h] - [a, \sigma(t)], [0, h - \mu(t)]]$$
$$= D[[0, h - \mu(t)], [0, h - \mu(t)]]$$
$$= 0.$$

Hence,

$$D[F(t + h) -_H F(\sigma(t)), [0, 1](h - \mu(t))] \le \varepsilon |\mu(t) - h| \quad \text{for all} \quad h \ge 0.$$

Similarly, for all $h \ge 0$, we have

$$D[F(\sigma(t)) -_H F(t - h), [0, 1](\mu(t) + h)] = 0 \le \varepsilon |\mu(t) + h|.$$

The following definition of Δ_g-differentiability, analogous to that of g-differentiability in Chalco-Cano *et al.* (2011), can be regarded as an improvement of Δ_H-differentiability.

Definition 2.5. Assume that $F \colon \mathbb{T} \to K_c^n$ is a set-valued function and $t \in \mathbb{T}^\kappa$. Let $\Delta_g F(t)$ be an element of K_c^n (provided it exists) with the property that, for given any $\varepsilon > 0$, there exists a neighborhood $U_\mathbb{T}$ of t (i.e., $U_\mathbb{T} = (t - \delta, t + \delta) \cap \mathbb{T}$ for some $\delta > 0$) such that

$$D[F(t+h) -_g F(\sigma(t)), \Delta_g F(t)(h - \mu(t))] \leq \varepsilon |h - \mu(t)| \qquad (2.2)$$

for all $t + h \in U_\mathbb{T}$ with $|h| < \delta$. Then $\Delta_g F(t)$ is said to be the Δ_g-derivative of F at t.

We say that F is Δ_g-differentiable at t provided its Δ_g-derivative exists at t. Moreover, we say F is Δ_g-differentiable on $\mathbb{T}$ provided its Δ_g-derivative exists at each $t \in \mathbb{T}^\kappa$. The set-valued function $\Delta_g F \colon \mathbb{T} \to K_c^n$ is then called the Δ_g-derivative of F on $\mathbb{T}$.

Let $\mathbf{D} \subset \mathbb{T}$. We denote the sets of all rd-continuous set-valued functions $F \colon \mathbf{D} \to K_c^n$ by

$$\mathcal{C}_{\mathrm{rd}} = \mathcal{C}_{\mathrm{rd}}(\mathbf{D}) = \mathcal{C}_{\mathrm{rd}}(\mathbf{D}, K_c^n),$$

and set of all Δ_g-differentiable set-valued functions $F \colon \mathbf{D} \to K_c^n$ whose Δ_g-derivative is rd-continuous by

$$\mathcal{C}_{\mathrm{rd}}^1 = \mathcal{C}_{\mathrm{rd}}^1(\mathbf{D}) = \mathcal{C}_{\mathrm{rd}}^1(\mathbf{D}, K_c^n).$$

It is significant to note that if we restrict ourselves to single-valued mappings, then the previous notions reduce to their ordinary counterparts, i.e., to the usual rd-continuity and delta differentiability in $\mathbb{T}$, see Bohner and Peterson (2001).

Proposition 2.3. *Assume that $F \colon \mathbb{T} \to K_c^n$ is a set-valued function and $t \in \mathbb{T}$. Then we have the following:*

(I) *If F is Δ_g-differentiable at t, then F is continuous at t.*
(II) *If F is continuous at t and t is right-scattered, then F is Δ_g-differentiable at t. Moreover, we have*

$$\Delta_g F(t) = \frac{F(\sigma(t)) -_g F(t)}{\mu(t)}.$$

(III) *If t is right-dense, then F is Δ_g-differentiable at t if and only if*

$$\Delta_g F(t) = \lim_{h \to 0} \frac{F(t+h) -_g F(t)}{h}. \tag{2.3}$$

(IV) *If F is Δ_g-differentiable at t, then*

$$F(\sigma(t)) = F(t) + \mu(t)\Delta_g F(t).$$

Proof. Part (I). Assume that F is Δ_g-differentiable at t. Let $\varepsilon \in (0,1)$. Define

$$\varepsilon^* = \varepsilon[\|\Delta_g F(t)\| + 1]^{-1}.$$

Clearly, $\varepsilon^* \in (0,1)$. Here, we define

$$D[X, \theta] = \|X\| = \sup_{x \in X} \|x\|$$

with $\theta = \{0\}$. It is easy to verify that, for $X, Y \in K_c^n$, if the g-difference of X and Y exists, then

$$D[X, Y] \le \|X -_g Y\|.$$

By Definition 2.5, there exists a neighborhood $U_{\mathbb{T}}$ of t such that

$$D[F(t+h) -_g F(\sigma(t)), \Delta_g F(t)(h - \mu(t))] \le \varepsilon^* |h - \mu(t)|$$

for all $t + h \in U_{\mathbb{T}}$. Therefore, for all $t + h \in U_{\mathbb{T}} \cap (t - \varepsilon^*, t + \varepsilon^*)$, we have

$$
\begin{aligned}
D[F(t+h), F(t)] &\le \|F(t+h) -_g F(t)\| = D[F(t+h) -_g F(t), \theta] \\
&= D[(F(t+h) -_g F(\sigma(t))) + (F(\sigma(t)) -_g F(t)), \\
&\qquad \Delta_g F(t)(h - \mu(t)) + \Delta_g F(t)\mu(t) - h\Delta_g F(t)] \\
&\le D[F(t+h) -_g F(\sigma(t)), \Delta_g F(t)(h - \mu(t))] \\
&\qquad + D[F(\sigma(t)) -_g F(t), \Delta_g F(t)\mu(t)] + D[h\Delta_g F(t), \theta] \\
&\le \varepsilon^* |h - \mu(t)| + \varepsilon^* \mu(t) + h\|\Delta_g F(t)\| \\
&\le \varepsilon^* h + h\|\Delta_g F(t)\| \\
&< \varepsilon^* (1 + \|\Delta_g F(t)\|).
\end{aligned}
$$

That is, for all $t + h \in U_{\mathbb{T}} \cap (t - \varepsilon^*, t + \varepsilon^*)$,

$$D[F(t+h), F(t)] < \varepsilon.$$

Similarly, one has

$$D[F(t), F(t-h)] \leq \varepsilon.$$

It follows that F is continuous at t.

Part (II). Assume that F is continuous at t and t is right-scattered. By the continuity of F, we have

$$\lim_{h \to 0^+} \frac{F(t+h) -_g F(\sigma(t))}{h - \mu(t)} = \frac{F(t) -_g F(\sigma(t))}{t - \sigma(t)} = \frac{F(t) -_g F(\sigma(t))}{-\mu(t)}$$

and

$$\lim_{h \to 0^+} \frac{F(\sigma(t)) -_g F(t-h)}{h + \mu(t)} = \frac{F(\sigma(t)) -_g F(t)}{\sigma(t) - t} = \frac{F(\sigma(t)) -_g F(t)}{\mu(t)}.$$

This implies that

$$D\left[\frac{F(\sigma(t)) -_g F(t)}{\mu(t)}, \frac{F(t) -_g F(\sigma(t))}{-\mu(t)} \right]$$

$$\leq \left\| \frac{F(\sigma(t)) -_g F(t)}{\mu(t)} -_g \frac{F(t) -_g F(\sigma(t))}{-\mu(t)} \right\|$$

$$= 0.$$

Hence,

$$\frac{F(t) -_g F(\sigma(t))}{-\mu(t)} = \frac{F(\sigma(t)) -_g F(t)}{\mu(t)}.$$

Moreover, for given $\varepsilon > 0$, there exists a neighborhood $U_{\mathbb{T}}$ of t such that

$$D\left[\frac{F(t+h) -_g F(\sigma(t))}{h - \mu(t)}, \frac{F(\sigma(t)) -_g F(t)}{\mu(t)} \right] \leq \varepsilon,$$

for all $t + h \in U_{\mathbb{T}}$. It follows that

$$D\left[F(t+h) -_g F(\sigma(t)), \frac{F(\sigma(t)) -_g F(t)}{\mu(t)}(h - \mu(t)) \right] \leq \varepsilon |h - \mu(t)|$$

for all $t + h \in U_{\mathbb{T}}$. Now, in the virtue of Definition 2.5, we get that F is Δ_g-differentiable at t and

$$\Delta_g F(t) = \frac{F(\sigma(t)) -_g F(t)}{\mu(t)}$$

as desired.

Part (III). Assume that F is Δ_g-differentiable at t and t is right-dense. Let $\varepsilon > 0$ be given. Since F is Δ_g-differentiable at t, there exists a neighborhood $U_{\mathbb{T}}$ of t such that

$$D[F(t + h) -_g F(\sigma(t)), \Delta_g F(t)(h - \mu(t))] \leq \varepsilon |h - \mu(t)|$$

for all $t + h \in U_{\mathbb{T}}$. Since $\sigma(t) = t$, i.e., $\mu(t) = 0$, we have that

$$D[F(t + h) -_g F(t), h\Delta_g F(t)] \leq \varepsilon h$$

and

$$D[F(t) -_g F(t - h), h\Delta_g F(t)] \leq \varepsilon h$$

for all $t + h \in U_{\mathbb{T}}$. This yields

$$D\left[\frac{F(t + h) -_g F(t)}{h}, \Delta_g F(t) \right] \leq \varepsilon$$

for all $t + h \in U_{\mathbb{T}}$. Therefore, from the arbitrariness of ε, we get that

$$\lim_{h \to 0} \frac{F(t + h) -_g F(t)}{h} = \Delta_g F(t).$$

Conversely, assume that t is right-dense and (2.3) hold. Then, it follows that, for any given $\varepsilon > 0$, there is a neighborhood $U_{\mathbb{T}}$ of t such that

$$D\left[\frac{F(t + h) -_g F(t)}{h}, \Delta_g F(t) \right] \leq \varepsilon$$

for all $t + h \in U_{\mathbb{T}}$. By means of this inequality, we immediately obtain the desired result.

Part (IV). If $\sigma(t) = t$, then $\mu(t) = 0$ and we have that

$$F(\sigma(t)) = F(t)$$
$$= F(t) + \mu(t)\Delta_g F(t).$$

Otherwise, $\sigma(t) > t$, then, by (II), we have

$$F(\sigma(t)) = F(t) + \mu(t)\frac{F(\sigma(t)) -_g F(t)}{\mu(t)}$$
$$= F(t) + \mu(t)\Delta_g F(t).$$

This guarantees that Part (IV) is true and completes the proof. □

Remark 2.1. We consider the two standard cases $\mathbb{T} = \mathbb{R}$ and $\mathbb{T} = \mathbb{Z}$:

(1) If $\mathbb{T} = \mathbb{R}$, then Proposition 2.3(III) yields that $F \colon \mathbb{R} \to K_c^1$ is Δ_g-differentiable at $t \in \mathbb{R}$ if and only if

$$\Delta_g F(t) = \lim_{h \to 0} \frac{F(t + h) -_g F(t)}{h},$$

i.e., if and only if F is differentiable in the Hukuhara sense at t. In this case, we have

$$\Delta_g F(t) = \Delta_H F(t) \quad \text{for} \quad t \in \mathbb{R}.$$

(2) If $\mathbb{T} = \mathbb{Z}$, then Proposition 2.3(II) yields that $F \colon \mathbb{Z} \to K_c^1$ is Δ_g-differentiable at $t \in \mathbb{Z}$ with

$$\Delta_g F(t) = \frac{F(\sigma(t)) -_g F(t)}{\mu(t)}$$
$$= F(t + 1) -_g F(t)$$
$$= \Delta F(t),$$

where Δ_g is the usual forward set-valued difference operator.

(3) It is also important to note that if we restrict ourselves to single-valued mappings, then the previous notions reduce to their classical counterparts, i.e., to ordinary continuity and differentiability in $\mathbb{R}$ when $\mathbb{T} = \mathbb{R}$ and to ordinary forward difference operators in $\mathbb{Z}$ when $\mathbb{T} = \mathbb{Z}$.

Now, we are able to find the Δ_g-derivatives of sum and scalar products of Δ_g-differentiable functions in the semilinear metric space K_c^n. This is possible according to the following proposition.

Proposition 2.4. *Assume that $F, G\colon \mathbb{T} \to K_c^n$ are Δ_g-differentiable. Then we have the following:*

(d1) *The sum $F + G$ defined by*

$$(F + G)(t) = F(t) + G(t) = \{x + y \colon x \in F(t),\ y \in G(t)\},$$

for each $t \in \mathbb{T}$, is Δ_g-differentiable at $t \in \mathbb{T}^\kappa$. Moreover,

$$\Delta_g(F + G)(t) = \Delta_g F(t) + \Delta_g G(t) \quad \text{for} \quad t \in \mathbb{T}^\kappa.$$

(d2) *The difference $F -_g G$ defined by*

$$(F -_g G)(t) = F(t) -_g G(t)$$

for each $t \in \mathbb{T}$, is Δ_g-differentiable at $t \in \mathbb{T}^\kappa$. Moreover,

$$\Delta_g(F -_g G)(t) = \Delta_g F(t) -_g \Delta_g G(t) \quad \text{for} \quad t \in \mathbb{T}^\kappa.$$

(d3) *For any constant λ, λF is Δ_g-differentiable at $t \in \mathbb{T}^\kappa$ with*

$$\Delta_g(\lambda F)(t) = \lambda \Delta_g F(t) \quad \text{for} \quad t \in \mathbb{T}^\kappa.$$

(d4) *The product function FG defined by*

$$(FG)(t) = F(t)G(t)$$

for each $t \in \mathbb{T}$, is Δ_g-differentiable at $t \in \mathbb{T}^\kappa$ with

$$\begin{aligned}
\Delta_g(FG)(t) &= F(\sigma(t))\Delta_g G(t) + G(t)\Delta_g F(t) \\
&= F(t)\Delta_g G(t) + G(\sigma(t))\Delta_g F(t) \quad \text{for} \quad t \in \mathbb{T}^\kappa.
\end{aligned}$$

Proof. (d1). Since F and G are Δ_g-differentiable at $t \in \mathbb{T}^\kappa$, for any given $\varepsilon > 0$, there exist neighborhoods $U_\mathbb{T}$ and $V_\mathbb{T}$ of t such that

$$D[F(t + h) -_g F(\sigma(t)), \Delta_g F(t)(h - \mu(t))] \leq \frac{\varepsilon}{2}|h - \mu(t)|$$

for all $t + h \in U_\mathbb{T}$ and

$$D[G(t + h) -_g G(\sigma(t)), \Delta_g G(t)(h - \mu(t))] \leq \frac{\varepsilon}{2}|h - \mu(t)|$$

for all $t + h \in V_\mathbb{T}$. Then, for all $t + h \in U_\mathbb{T} \cap V_\mathbb{T} := W_\mathbb{T}$, we have

$$D[(F + G)(t + h) -_g (F + G)(\sigma(t)), (\Delta_g F(t) + \Delta_g G(t))(h - \mu(t))]$$
$$= D[F(t + h) -_g F(\sigma(t)) + G(t + h) -_g G(\sigma(t)),$$
$$\Delta_g F(t)(h - \mu(t)) + \Delta_g G(t)(h - \mu(t))]$$
$$\leq D[F(t + h) -_g F(\sigma(t)), \Delta_g F(t)(h - \mu(t))]$$
$$+ D[G(t + h) -_g G(\sigma(t)), \Delta_g G(t)(h - \mu(t))]$$
$$\leq \frac{\varepsilon}{2}|h - \mu(t)| + \frac{\varepsilon}{2}|h - \mu(t)|$$
$$= \varepsilon|h - \mu(t)|.$$

Therefore, $F + G$ is Δ_g-differentiable at t and

$$\Delta_g(F + G)(t) = \Delta_g F(t) + \Delta_g G(t) \quad \text{for} \quad t \in \mathbb{T}^\kappa,$$

as desired.

(d2). From Propositions 2.1(i) and (iv), it follows that

$$D[(F -_g G)(t + h) -_g (F -_g G)(\sigma(t)), (\Delta_g F(t) -_g \Delta_g G(t))(h - \mu(t))]$$
$$= D[F(t + h) -_g F(\sigma(t)) + [G(t) -_g G(\sigma(t + h)),$$
$$\Delta_g F(t)(h - \mu(t)) -_g \Delta_g G(t)(h - \mu(t))].$$

The remaining part of the proof is similar to that of (d1).

(d3). We assume that $\lambda > 0$. Otherwise, the desired result trivially holds. The differentiability of F at $t \in \mathbb{T}$ guarantees that there exists a neighborhood $U_{\mathbb{T}}$ of t such that, for any given $\varepsilon > 0$, we have

$$D[F(t+h)-_gF(\sigma(t)), \Delta_gF(t)(h-\mu(t))] \leq \frac{\varepsilon}{\lambda}|h-\mu(t)|$$

for all $t+h \in U_{\mathbb{T}}$. This implies that

$$\begin{aligned}
D[\lambda F(t+h)&-_g\lambda F(\sigma(t)), \lambda\Delta_gF(t)(h-\mu(t))] \\
&= \lambda D[F(t+h)-_gF(\sigma(t)), \Delta_gF(t)(h-\mu(t))] \\
&\leq \lambda\frac{\varepsilon}{\lambda}|h-\mu(t)| \\
&= \varepsilon|h-\mu(t)|.
\end{aligned}$$

for all $t+h \in U_{\mathbb{T}}$. Therefore, λF is Δ_g-differentiable at t and

$$\Delta_g(\lambda F)(t) = \lambda\Delta_gF(t) \quad \text{for} \quad t \in \mathbb{T}^\kappa,$$

as desired.

(d4). Let $\varepsilon \in (0,1)$. Define $\varepsilon_1 = \varepsilon(1 + \|G(t)\| + \|F(\sigma(t))\| + \|\Delta_gF(t)\|)^{-1}$. Then, $\varepsilon_1 \in (0,1)$ and there exist neighborhoods $U_{\mathbb{T}}^1$, $U_{\mathbb{T}}^2$, and $U_{\mathbb{T}}^3$ of t such that

$$D[F(t+h)-_gF(\sigma(t)), \Delta_gF(t)(h-\mu(t))] \leq \varepsilon_1|h-\mu(t)|$$

for all $t+h \in U_{\mathbb{T}}^1$, and

$$D[G(t+h)-_gG(\sigma(t)), \Delta_gG(t)(h-\mu(t))] \leq \varepsilon_1|h-\mu(t)|$$

for all $t+h \in U_{\mathbb{T}}^2$. From Proposition 2.3(I), it follows that

$$D[G(t+h), G(t)] \leq \varepsilon_1 \quad \text{and} \quad \|G(t+h)\| \leq \|G(t)\| + 1$$

for all $t+h \in U_{\mathbb{T}}^3$. Let $U_{\mathbb{T}} := U_{\mathbb{T}}^1 \cap U_{\mathbb{T}}^2 \cap U_{\mathbb{T}}^3$ and $t+h \in U_{\mathbb{T}}$. Then,

$$\begin{aligned}
D[(FG)(t+h)&-_g(FG)(\sigma(t)), (F(\sigma(t))\Delta_gG(t) \\
&+ G(t)\Delta_gF(t))(h-\mu(t))] \\
\leq\ & D[(G(t+h)-_gG(\sigma(t)))F(\sigma(t)), F(\sigma(t))\Delta_gG(t)(h-\mu(t))] \\
&+ D[(F(t+h)-_gF(\sigma(t)))G(t+h), G(t+h)\Delta_gF(t)(h-\mu(t))] \\
&+ D[\theta, G(t)\Delta_gF(t)(h-\mu(t)) - G(t+h)\Delta_gF(t)(h-\mu(t))]
\end{aligned}$$

$$\leq D[G(t+h)-_gG(\sigma(t)), \Delta_gG(t)(h-\mu(t))]\|F(\sigma(t))\|$$
$$+ D[F(t+h)-_gF(\sigma(t)), \Delta_gF(t)(h-\mu(t))]\|G(t+h)\|$$
$$+ D[G(t+h), G(t)]\|\Delta_gF(t)\|\|h-\mu(t)\|$$
$$\leq \varepsilon_1\|F(\sigma(t))\|(h-\mu(t)) + \varepsilon_1(\|G(t)\|+1)(h-\mu(t))$$
$$+ \varepsilon_1\|\Delta_gF(t)\|\|h-\mu(t)\|$$
$$= \varepsilon_1(1 + \|F(\sigma(t))\| + \|G(t)\| + \|\Delta_gF(t)\|)|h-\mu(t)|$$
$$= \varepsilon|h-\mu(t)|.$$

Thus,

$$\Delta_g(FG)(t) = F(\sigma(t))\Delta_gG(t) + G(t)\Delta_gF(t) \quad \text{for} \quad t \in \mathbb{T}^\kappa.$$

The second product rule follows from this last equation by interchanging the multivalued functions F and G. This completes the proof. $\square$

2.3.2 *Integral of set-valued functions*

In what follows, we introduce the concept of integral of set-valued functions. By $\mathcal{S}_F(\mathbf{D})$, we mean the set of all delta integrable selectors of F on $\mathbf{D}$.

Definition 2.6. Let $\mathbf{D} \subset \mathbb{T}$. A set-valued function $F: \mathbb{T} \to K_c^n$ is called Δ_g-integrable on $\mathbf{D}$ provided F has at least one Δ-integrable selector on $\mathbf{D}$. In this case, we define the Δ_g-integral of F on $\mathbf{D}$, denoted by

$$\int_{\mathbf{D}} F(s)\Delta_gs,$$

as the set

$$\int_{\mathbf{D}} F(s)\Delta_gs = \left\{ \int_{\mathbf{D}} f(s)\Delta s : f \in S_F(\mathbf{D}) \right\}.$$

Example 2.3. Let $\mathbf{D} \subset \mathbb{T}$ be an interval, $f_1, f_2 \in \mathcal{C}_{\mathrm{rd}}(\mathbf{D}, \mathbb{R})$ with $f_1(t) < f_2(t)$ for all $t \in \mathbf{D}$, and a set-valued function

$$F = \{f \in \mathcal{C}_{\mathrm{rd}}(\mathbf{D}, \mathbb{R}) : f_1(t) \le f(t) \le f_2(t) \quad \text{for all } t \in \mathbf{D}\}.$$

Thus, $\mathcal{S}_F(\mathbf{D}) = F$ and

$$\int_{\mathbf{D}} F(s) \Delta_g s = \left\{ \int_{\mathbf{D}} f(s) \Delta s : f \in F \right\} \subset \left[\int_{\mathbf{D}} f_1(s) \Delta s, \int_{\mathbf{D}} f_2(s) \Delta s \right].$$

Proposition 2.5. *Assume that $t_0, T \in \mathbb{T}$ and $F, G \colon [t_0, T]_{\mathbb{T}} \to K_c^n$ are Δ_g-integrable. Then we have the following:*

(i1) $\displaystyle \int_{t_0}^{T} [F(s) \pm_g G(s)] \Delta_g s = \int_{t_0}^{T} F(s) \Delta_g s \pm_g \int_{t_0}^{T} G(s) \Delta_g s.$

(i2) $\displaystyle \int_{t_0}^{t} \lambda F(s) \Delta_g s = \lambda \int_{t_0}^{t} F(s) \Delta_g s \quad \text{for} \quad \lambda \in \mathbb{R}, \quad t \in [t_0, T]_{\mathbb{T}}.$

(i3) $\displaystyle \int_{t_0}^{T} F(s) \Delta_g s = \int_{t_0}^{t} F(s) \Delta_g s + \int_{t}^{T} F(s) \Delta_g s \quad \text{for} \quad t \in [t_0, T]_{\mathbb{T}}.$

(i4) $\displaystyle \int_{t}^{t} F(s) \Delta_g s = \theta \text{ for } t \in [t_0, T]_{\mathbb{T}}.$

(i5) *If $F \in \mathcal{C}_{\mathrm{rd}}(\mathbb{T}, K_c^n)$, then, for any $t_1, t_2 \in \mathbb{T}$, we have*

$$\int_{t_1}^{t_2} F(s) \Delta_g s = \theta -_g \int_{t_2}^{t_1} F(s) \Delta_g s.$$

(i6) *If $f \in \mathcal{S}_F([t_0, T]_{\mathbb{T}})$ and $g \in \mathcal{S}_G([t_0, T]_{\mathbb{T}})$ imply that $f, g \in \mathcal{C}_{\mathrm{rd}}([t_0, T]_{\mathbb{T}}, \mathbb{R})$, then $D[F(\cdot), G(\cdot)] \colon [t_0, T]_{\mathbb{T}} \to \mathbb{R}^+$ is delta integrable and*

$$D \left[\int_{t_0}^{T} F(s) \Delta_g s, \int_{t_0}^{T} G(s) \Delta_g s \right] \le \int_{t_0}^{T} D[F(s), G(s)] \Delta s.$$

(i7) *If $f \in \mathcal{S}_F([t_0, T]_{\mathbb{T}})$ implies that $f \in \mathcal{C}_{\mathrm{rd}}([t_0, T]_{\mathbb{T}}, \mathbb{R})$, then $\|F(\cdot)\| \colon [t_0, T]_{\mathbb{T}} \to \mathbb{R}^+$ is delta integrable and*

$$\left\| \int_{t_0}^{T} F(s) \Delta_g s \right\| \le \int_{t_0}^{T} \|F(s)\| \Delta s.$$

Proof. (i1). For any

$$u \in \int_{t_0}^{T} [F(s) + G(s)] \Delta_g s,$$

in view of Theorem 1.5, there exist $f \in \mathcal{S}_F([t_0, T]_\mathbb{T})$ and $g \in \mathcal{S}_G([t_0, T]_\mathbb{T})$ such that

$$u = \int_{t_0}^{T} [f(s) + g(s)] \Delta s$$

$$= \int_{t_0}^{T} f(s) \Delta s + \int_{t_0}^{T} g(s) \Delta s$$

$$\in \int_{t_0}^{T} F(s) \Delta_g s + \int_{t_0}^{T} G(s) \Delta_g s.$$

This yields that

$$\int_{t_0}^{T} [F(s) + G(s)] \Delta_g s \subset \int_{t_0}^{T} F(s) \Delta_g s + \int_{t_0}^{T} G(s) \Delta_g s.$$

By an analogous process, we can obtain the converse inclusion. Therefore, (i1) is true.

(i2) It is trivial.

(i3) By means of Theorem 1.5, for any $f \in \mathcal{S}_F([t_0, T]_\mathbb{T})$, we have

$$\int_{t_0}^{T} f(s) \Delta s = \int_{t_0}^{t} f(s) \Delta s + \int_{t}^{T} f(s) \Delta s$$

which implies that

$$\int_{t_0}^{T} F(s) \Delta_g s \subset \left(\int_{t_0}^{t} F(s) \Delta_g s + \int_{t}^{T} F(s) \Delta_g s \right). \qquad (2.4)$$

On the other hand, taking

$$u \in \int_{t_0}^{t} F(s) \Delta_g s + \int_{t}^{T} F(s) \Delta_g s,$$

we can find $f_1 \in \mathcal{S}_F([t_0, t]_\mathbb{T})$, $f_2 \in \mathcal{S}_F([t, T]_\mathbb{T})$ such that

$$u = \int_{t_0}^{t} f_1(s) \Delta s + \int_{t}^{T} f_2(s) \Delta s.$$

Now, let

$$\widetilde{f_1}(s) = \begin{cases} f_1(s) & \text{for } s \in [t_0, t]_{\mathbb{T}} \\ 0 & \text{for } s \in [t, T]_{\mathbb{T}}, \end{cases}$$

and

$$\widetilde{f_2}(s) = \begin{cases} f_2(s) & \text{for } s \in [t, T]_{\mathbb{T}} \\ 0 & \text{for } s \in [t_0, t]_{\mathbb{T}}. \end{cases}$$

Clearly, $\widetilde{f_i} \in \mathcal{S}_F([t_0, T]_{\mathbb{T}})$ for $i = 1, 2$, and

$$\int_{t_0}^{T} \widetilde{f_1}(s)\Delta s = \int_{t_0}^{t} f_1(s)\Delta s \quad \text{and} \quad \int_{t_0}^{T} \widetilde{f_2}(s)\Delta s = \int_{t}^{T} f_2(s)\Delta s.$$

Hence,

$$u = \int_{t_0}^{T} \widetilde{f_1}(s)\Delta s + \int_{t_0}^{T} \widetilde{f_2}(s)\Delta s$$

$$= \int_{t_0}^{T} [\widetilde{f_1}(s) + \widetilde{f_2}(s)]\Delta s$$

$$\in \int_{t_0}^{T} F(s)\Delta_g s.$$

This shows that

$$\int_{t_0}^{t} F(s)\Delta_g s + \int_{t}^{T} F(s)\Delta_g s \subset \int_{t_0}^{T} F(s)\Delta_g s. \tag{2.5}$$

From (2.4) and (2.5), (i3) holds as desired.

(i4) In (i3), take $t = T$. Then we get (i4).

(i5) From (i3) and (i4), we have

$$\theta = \int_{t_1}^{t_1} F(s)\Delta_g s$$

$$= \int_{t_1}^{t_2} F(s)\Delta_g s + \int_{t_2}^{t_1} F(s)\Delta_g s.$$

Now, the definition of "$-_g$" implies that (i5) holds.

(i6) We first observe that, for given $f \in \mathcal{C}_{\text{rd}}([t_0, T]_{\mathbb{T}}, \mathbb{R})$, one has the inequalities

$$d(f(s), G(s)) \leq d(f(s), g(s))$$
$$\leq d(f(s), f(t)) + d(f(t), g(s))$$

for each $g(s) \in G(s)$. It follows that

$$d(f(s), G(s)) - d(f(s), f(t)) \leq \inf d(f(t), g(s))$$
$$= d(f(t), G(s))$$

so that

$$d(f(s), G(s)) - d(f(t), G(s)) \leq d(f(s), f(t)).$$

The same inequality holds with s and t interchanged, and thus, rd-continuity of the $d(f(\cdot), G(s))$ at $s \in [t_0, T]_\mathbb{T}$ follows for each $f \in \mathcal{C}_{\mathrm{rd}}([t_0, T]_\mathbb{T}, \mathbb{R})$. Moreover, we can similarly see the rd-continuity of the function $D[F(\cdot), G(\cdot)]$ on $[t_0, T]_\mathbb{T}$. This yields that the integral

$$\int_{t_0}^{T} D[(F(s), G(s)]\Delta s$$

is well defined. Finally, for given

$$u \in \int_{t_0}^{T} F(s)\Delta_g s,$$

there exists $f \in \mathcal{C}_{\mathrm{rd}}([t_0, T]_\mathbb{T}, \mathbb{R})$ such that

$$u = \int_{t_0}^{T} f(s)\Delta s.$$

Also, for any

$$v \in \int_{t_0}^{T} G(s)\Delta_g s,$$

there exists $g \in \mathcal{C}_{\mathrm{rd}}([t_0, T]_\mathbb{T}, \mathbb{R})$ such that

$$v = \int_{t_0}^{T} g(s)\Delta s.$$

Since

$$d(u, v) = \left| \int_{t_0}^{T} f(s) \Delta s - \int_{t_0}^{T} g(s) \Delta s \right|$$

$$\leq \int_{t_0}^{T} |f(s) - g(s)| \Delta s$$

holds for all $g \in \mathcal{S}_G([t_0, T]_{\mathbb{T}})$, combining the compactness of $F(s)$ and $G(s)$ for each $s \in ([t_0, T]_{\mathbb{T}})$, we assert that

$$d(u, v) \leq \int_{t_0}^{T} D[F(s), G(s)] \Delta s.$$

From the arbitrariness of u and v, it follows that (i6) holds, as desired. (i7) When we select $G(t) = \theta$ in (i6), we obtain the desired inequality. This completes the proof. $\qquad \square$

Definition 2.7. A set-valued function $F \colon \mathbb{T} \to K_c^n$ is called Δ_g pre-differentiable with (region of differentiation) **D** provided

(1) $\mathbf{D} \subset \mathbb{T}^{\kappa}$;

(2) $\mathbb{T}^{\kappa} \setminus \mathbf{D}$ is countable and contains no right-scattered elements of $\mathbb{T}$;

(3) F is Δ_g-differentiable at each $t \in \mathbf{D}$.

Remark 2.2. It is evident that F is regulated provided F is rd-continuous. A similar argument to Theorems 2.1 and 2.2, in addition, can be used to show that there exists a set-valued function $\mathfrak{F}$, which is Δ_g pre-differentiable with region of differentiation **D** such that $\Delta_H \mathfrak{F}(t) = F(t)$ for all $t \in \mathbf{D}$, for given regulated set-valued function F.

Definition 2.8. Let $G \colon \mathbb{T} \times \mathbb{T} \to K_c^n$ be a binary set-valued function. Then $G(t, s)$ is called uniformly Δ_g-differentiable with respect to t on $\mathbf{D} \subset \mathbb{T}$, provided there exists $\varepsilon > 0$ such that

$$D[G(t + h, s) -_g G(\sigma(t), s), \Delta_g^t G(t, s)(h - \mu(t))] \leq \varepsilon |h - \mu(t)|$$

for each $s \in \mathbf{D}$, where $\Delta_g^t G(t, s)$ stands for the Δ_g-derivative of G with respect to first variable t.

Theorem 2.1. *Let $J \subset \mathbb{R}$ be an interval and $t_0, t \in J_{\mathbb{T}}$. For $F \in \mathcal{C}_{\mathrm{rd}}(J_{\mathbb{T}}, K_c^n)$, define a set-valued function $\mathfrak{F} \colon J_{\mathbb{T}} \to K_c^n$ by*

$$\mathfrak{F}(t) = X_0 + \int_{t_0}^{t} F(s) \Delta_g s, \quad \text{where } X_0 \in K_c^n.$$

Then we have the following:

(i) *The function $\mathfrak{F}$ is Δ_g-differentiable on $J_{\mathbb{T}}$ and we have the equality*

$$\Delta_g \mathfrak{F}(t) = F(t) \quad \text{for} \quad t \in J_{\mathbb{T}}.$$

(ii) *If F is rd-continuous at $t \in \mathbb{T}$, then*

$$\int_t^{\sigma(t)} F(s) \Delta_g s = \mu(t) F(t).$$

(iii) *Let $G \in \mathcal{C}_{\mathrm{rd}}^1(\mathbb{T} \times \mathbb{T}, K_c^n)$ be continuous at (t, t) and $G(t, s)$ be uniformly Δ_g-differentiable with respect to t on $\mathbb{T}$. If*

$$\mathfrak{F}(t) = \int_{t_0}^{t} G(t, s) \Delta_g s \quad \text{for} \quad t_0 \in \mathbb{T},$$

then

$$\Delta_g \mathfrak{F}(t) = G(\sigma(t), t) + \int_{t_0}^{t} \Delta_g^t G(t, s) \Delta_g s \quad \text{for} \quad t \in J_{\mathbb{T}},$$

where $\Delta_g^t G(t, s)$ stands for the Δ_g-derivative of G with respect to first variable t.

Proof. (i) Suppose that the region of differentiation of F is $\mathbf{D}$ and $t \in \mathbf{D}$. Let $n \in \mathbb{N}$. Consider the statement

$$S(t) \colon \begin{cases} \text{there exists a } \Delta_g \text{ pre-differentiable } (\mathfrak{F}_{nt}, \mathbf{D}_{nt}), \mathfrak{F}_{nt} \colon [t_0, t]_{\mathbb{T}} \to \mathbb{R} \\ \text{with } \mathfrak{F}_{nt}(t_0) = X_0 \text{ and } D[\Delta_g \mathfrak{F}_{nt}(s), F(s)] \le \frac{1}{n} \text{ for } s \in \mathbf{D}_{nt}. \end{cases}$$

We use the time scale version of induction principle given in Theorem 1.7 (Bohner and Peterson, 2001):

I. Let $t = t_0$, $\mathbf{D}_{nt} = \emptyset$, and $\mathfrak{F}_{nt}(t_0) = X_0$. Then the statement $S(t_0)$ is true.

II. Let t be right-scattered in $J_{\mathbb{T}}$ and assume that $S(t)$ is true. Define

$$\mathbf{D}_{n\sigma(t)} = \mathbf{D}_{nt} \cup \{t\},$$

and $\mathfrak{F}_{n\sigma(t)}$ on $[t_0, \sigma(t)]_{\mathbb{T}}$ by

$$\mathfrak{F}_{n\sigma(t)}(s) = \begin{cases} \mathfrak{F}_{n\sigma(t)}(s) & \text{for} \quad s \in [t_0, t]_{\mathbb{T}} \\ \mathfrak{F}_{n\sigma(t)}(t) + \mu(t)F(t) & \text{for} \quad s = \sigma(t). \end{cases}$$

Then,

$$\mathfrak{F}_{n\sigma(t)}(t_0) = \mathfrak{F}_{nt}(t_0) = X_0,$$

$$D[\Delta_g \mathfrak{F}_{n\sigma(t)}(s), F(s)] = D[\Delta_g \mathfrak{F}_{nt}(s), F(s)] \leq \frac{1}{n} \quad \text{for} \quad s \in \mathbf{D}_{nt},$$

and

$$D\left[\frac{\mathfrak{F}_{n\sigma(t)}(\sigma(t)) -_g \mathfrak{F}_{n\sigma(t)}(t)}{\mu(t)}, F(t)\right]$$

$$= D\left[\frac{\mathfrak{F}_{n\sigma(t)}(t) + \mu(t)F(t) -_g \mathfrak{F}_{n\sigma(t)}(t)}{\mu(t)}, F(t)\right]$$

$$= D\left[\frac{\mu(t)F(t)}{\mu(t)}, F(t)\right]$$

$$= \theta$$

$$\leq \frac{1}{n},$$

and therefore the statement $S(\sigma(t))$ is true.

III. Let t be a right-dense point in $J_{\mathbb{T}}$ and assume that $S(t)$ is true. Since F is rd-continuous, it is regulated. Since t is right-dense, it follows that

$$F(t^+) = \lim_{s \to t, \, s > t} F(s) \quad \text{exists.}$$

Hence, there exists a neighborhood $U_{\mathbb{T}}$ of $t \in J_{\mathbb{T}}$ with

$$D[F(s), F(t^+)] \leq \frac{1}{n} \quad \text{for} \quad s \in U_{\mathbb{T}} \cap (t, \infty). \tag{2.6}$$

Let $r \in U_{\mathbb{T}} \cap (t, \infty)$. Define

$$\mathbf{D}_n = (\mathbf{D}_{nt} \setminus \{t\}) \cup [t, r]_{\mathbb{T}}^{\kappa},$$

and $\mathfrak{F}_{nr}$ on $[t_0, r]_{\mathbb{T}}$ by

$$\mathfrak{F}_{nr}(s) = \begin{cases} \mathfrak{F}_{nt}(s) & \text{for} \quad s \in [t_0, t]_{\mathbb{T}} \\ \mathfrak{F}_{nt}(t) + (s - t)F(t^+) & \text{for} \quad s \in (t, r]_{\mathbb{T}}. \end{cases}$$

Then, $\mathfrak{F}_{nr}$ is continuous at t and hence on $[t_0, r]_{\mathbb{T}}$. Further, $\mathfrak{F}_{nr}$ is differentiable on $(t, r]_{\mathbb{T}}^{\kappa}$ with

$$\Delta_g \mathfrak{F}_{nr}(s) = F(t^+) \quad \text{for} \quad s \in (t, r]_{\mathbb{T}}^{\kappa}.$$

Hence, $\mathfrak{F}_{nr}$ is Δ_g pre-differentiable with $[t_0, t)_{\mathbb{T}}$. Since t is right-dense, it follows that $\mathfrak{F}_{nr}$ is Δ_g pre-differentiable with $\mathbf{D}_{nr}$. Keeping in mind $S(t)$ and (2.6), we obtain

$$D\left[\Delta_g \mathfrak{F}_{nr}(s), F(s)\right] \le \frac{1}{n} \quad \text{for} \quad s \in \mathbf{D}_{nr}.$$

Therefore, the statement $S(r)$ is true for all $r \in U_{\mathbb{T}} \cap (t, \infty)_{\mathbb{T}}$.

IV. Now, let t be left-dense and suppose that $S(r)$ is true for $r < t$. Since F is rd-continuous, it is regulated and so

$$F(t^-) = \lim_{s \to t, \, s < t} F(s) \quad \text{exists.}$$

Hence, there exists a neighborhood $U_{\mathbb{T}}$ of t with

$$D[F(s), F(t^-)] \le \frac{1}{n} \quad \text{for} \quad s \in U_{\mathbb{T}} \cap (-\infty, t)_{\mathbb{T}}.$$

Fix some $r \in U_{\mathbb{T}} \cap (t, \infty)_{\mathbb{T}}$ and define

$$\mathbf{D}_{nt} = \begin{cases} \mathbf{D}_{nr} \cup (r, t)_{\mathbb{T}} & \text{if } r \text{ is right-dense} \\ \mathbf{D}_{nr} \cup [r, t)_{\mathbb{T}} & \text{if } r \text{ is right-scattered,} \end{cases}$$

and $\mathfrak{F}_{nt}$ on $[t_0, t]_{\mathbb{T}}$ by

$$\mathfrak{F}_{nt}(s) = \begin{cases} \mathfrak{F}_{nt}(s) & \text{if } s \in (t_0, r]_{\mathbb{T}} \\ \mathfrak{F}_{nt}(t) + (s - t)F(t^-) & \text{if } s \in (r, t]_{\mathbb{T}}. \end{cases}$$

Note that $\mathfrak{F}_{nt}$ is continuous at r and hence on $[t_0, t]_{\mathbb{T}}$. Since

$$\Delta_g \mathfrak{F}_{nt}(s) = F(t^-) \quad \text{for} \quad s \in (r, t]_{\mathbb{T}},$$

$\mathfrak{F}_{nt}$ is Δ_g pre-differentiable with $\mathbf{D}_{nt}$ and

$$D[\Delta_g \mathfrak{F}_{nt}(s), F(s)] \leq \frac{1}{n} \quad \text{for} \quad s \in \mathbf{D}_{nt}.$$

Hence, the statement $S(t)$ is true. By the induction principle, it follows that $S(t)$ is true for all $t \geq t_0$, $t \in \mathbb{T}$. Similarly, we can show that $S(t)$ valid for $t \leq t_0$. Hence, $\mathfrak{F}_n$ is Δ_g pre-differentiable with $\mathbf{D}_n$, $\mathfrak{F}_n(t_0) = X_0$, and

$$D[\Delta_g \mathfrak{F}_n(t), F(t)] \leq \frac{1}{n} \quad \text{for} \quad t \in \mathbf{D}_n.$$

Now, let

$$\mathfrak{F} = \lim_{n \to \infty} \mathfrak{F}_n \quad \text{and} \quad \mathbf{D} = \bigcap_{n \in \mathbb{N}} \mathbf{D}_n.$$

Then, $\mathfrak{F}_n(t_0) = X_0$, $\mathfrak{F}$ is Δ_g pre-differentiable with $\mathbf{D}$, and

$$\Delta_g \mathfrak{F}(t) = \lim_{n \to \infty} \mathfrak{F}_n(t) = F(t) \quad \text{for} \quad t \in \mathbf{D}.$$

Next, if $t \in U_{\mathbb{T}} \subset J_{\mathbb{T}}$, then t is a right-dense point of $\mathbb{T}$. For any $t + h \in U_{\mathbb{T}}$, we have

$$\mathfrak{F}(t + h) -_g \mathfrak{F}(\sigma(t)) = \int_{t_0}^{t+h} F(s) \Delta_g s -_g \int_{t_0}^{\sigma(t)} F(s) \Delta_g s$$

$$= \int_{\sigma(t)}^{t+h} F(s) \Delta_g s.$$

So,

$$D[\mathfrak{F}(t + h) -_g \mathfrak{F}(\sigma(t)), F(t)(h - \mu(t))]$$

$$= D\left[\int_{\sigma(t)}^{t+h} F(s) \Delta_g s, \int_{\sigma(t)}^{t+h} F(t) \Delta_g s\right]$$

$$\leq \int_{\sigma(t)}^{t+h} D[F(s), F(t)] \Delta s.$$

Since t is right-dense, the limits on the right side of the above two expressions will approach to zero as h goes to zero. This implies that $\Delta_g \mathfrak{F}(t)$ exists and

$$\Delta_g \mathfrak{F}(t) = F(t) \quad \text{on} \quad J_{\mathbb{T}}.$$

(ii) Part (i) guarantees the existence of Δ_g-derivative of $\mathfrak{F}$, and

$$\int_t^{\sigma(t)} F(s) \Delta_g s = \mathfrak{F}(\sigma(t)) - \mathfrak{F}(t)$$

$$= \mu(t) \Delta_g \mathfrak{F}(t)$$

$$= \mu(t) F(t).$$

(iii) Let $t \in \mathbb{T}$ be given and

$$\bar{G}(t, s) = G(\sigma(t), t) + \int_{t_0}^t \Delta_g^t G(t, s) \Delta_g s.$$

For any $\varepsilon > 0$, the uniformly Δ_g-differentiability guarantees that there exists a neighborhood $U_{\mathbb{T}}^1$ of t such that

$$D[G(t + h, s) -_g G(\sigma(t), s), \Delta_g^t G(t, s)(h - \mu(t))] \leq \frac{\varepsilon}{3|t + 1 - t_0|}$$

for all $s \in \mathbb{T}$. By the continuity of G at (t, t), there exists a neighborhood $U_{\mathbb{T}}^2$ of t such that

$$D[G(\sigma(t), s), G(\sigma(t), t)] \leq \frac{\varepsilon}{3|h - \mu(t)|} \quad \text{for} \quad s \in U_{\mathbb{T}}^2.$$

Since $\Delta_g^t G(t, s)$ is regulated, by Lemma 2.2, there exists a positive number M such that

$$\|\Delta_g^t G(t, s)\| \leq M.$$

Let $|h|$ be sufficient small such that $t + h \in U_{\mathbb{T}}^1 \cap U_{\mathbb{T}}^2 := U_{\mathbb{T}}$ and

$$|h| < \min \left\{ 1, \frac{\varepsilon}{3M} \right\}.$$

Then,

$$D\left[\mathfrak{F}(t+h)-_g\mathfrak{F}(\sigma(t)), \bar{G}(t,s)(h-\mu(t))\right]$$

$$= D\left[\int_{t_0}^{t+h} G(t+h,s)\Delta_g s -_g \int_{t_0}^{\sigma(t)} G(\sigma(t),s)\Delta_g s, \bar{G}(t,s)(h-\mu(t))\right]$$

$$= D\left[\int_{t_0}^{t+h} (G(t+h,s)-_g G(\sigma(t),s))\Delta_g s + \int_{\sigma(t)}^{t+h} G(\sigma(t),s)\Delta_g s,\right.$$

$$(h-\mu(t))\left(G(\sigma(t),t) + \int_{t_0}^{t+h} \Delta_g{}^t G(t,s)\Delta_g s\right.$$

$$\left.\left. -_g \int_t^{t+h} \Delta_g^t G(t,s)\Delta_g s\right)\right]$$

$$\leq D\left[\int_{t_0}^{t+h} (G(t+h,s)-_g G(\sigma(t),s))\Delta_g s, (h-\mu(t))\right.$$

$$\left. \times \int_{t_0}^{t+h} \Delta_g^t G(t,s)\Delta_g s\right]$$

$$+ D\left[\int_{\sigma(t)}^{t+h} G(\sigma(t),s)\Delta_g s, (h-\mu(t))\right.$$

$$\left. \times \left(G(\sigma(t),t)-_g \int_t^{t+h} \Delta_g^t G(t,s)\Delta_g s\right)\right]$$

$$= D\left[\int_{t_0}^{t+h} (G(t+h,s)-_g G(\sigma(t),s))\Delta_g s, (h-\mu(t))\right.$$

$$\left. \times \int_{t_0}^{t+h} \Delta_g^t G(t,s)\Delta_g s\right]$$

$$+ D\left[\int_{\sigma(t)}^{t+h} G(\sigma(t),s)\Delta_g s, (h-\mu(t))G(\sigma(t),t)\right]$$

$$+ D\left[\theta, -_g(h-\mu(t))\int_t^{t+h} \Delta_g^t G(t,s)\Delta_g s\right]$$

$$\leq D \left[\int_{t_0}^{t+h} (G(t+h,s) -_g G(\sigma(t),s)) \Delta_g s, \, (h-\mu(t)) \right.$$

$$\left. \times \int_{t_0}^{t+h} \Delta_g^t G(t,s) \Delta_g s \right]$$

$$+ D \left[\int_{\sigma(t)}^{t+h} G(\sigma(t),s) \Delta_g s, \, \int_{\sigma(t)}^{t+h} G(\sigma(t),t) \Delta_g s \right]$$

$$+ \left\| (h-\mu(t)) \int_{t}^{t+h} \Delta_g^t G(t,s) \Delta_g s \right\|$$

$$\leq \int_{t_0}^{t+h} D \left[(G(t+h,s) -_g G(\sigma(t),s)), (h-\mu(t)) \Delta_g^t G(t,s) \right] \Delta s$$

$$+ \int_{\sigma(t)}^{t+h} D \left[G(\sigma(t),s), G(\sigma(t),t) \right] \Delta s + |h-\mu(t)|$$

$$\times \int_{t}^{t+h} \left\| \Delta_g^t G(t,s) \right\| \Delta s$$

$$\leq \int_{t_0}^{t+h} \frac{\varepsilon}{3|t+1-t_0|} \Delta s + \left| \int_{\sigma(t)}^{t+h} \frac{\varepsilon}{3|h-\mu(t)|} \Delta s \right| + M|h-\mu(t)|h$$

$$< \frac{|t+1-t_0|\varepsilon}{3|t+1-t_0|} + \frac{|h-\mu(t)|\varepsilon}{3|h-\mu(t)|} + \frac{\varepsilon}{3}.$$

Thus,

$$D \left[\mathfrak{F}(t+h) -_g \mathfrak{F}(\sigma(t)), \bar{G}(t,s)(h-\mu(t)) \right] \leq \varepsilon$$

for $t+h \in U_{\mathbb{T}}$. Consequently, (iii) holds. $\qquad\square$

Remark 2.3. A set-valued function $\mathfrak{F} \colon \mathbb{T} \to K_c^n$ is called a Δ_g-antiderivative of $F \colon \mathbb{T} \to K_c^n$ provided

$$\Delta_g \mathfrak{F}(t) = F(t) \quad \text{for all} \quad t \in \mathbb{T}.$$

Theorem 2.1 shows that every rd-continuous set-valued function F has a Δ_g-antiderivative $\mathfrak{F}$ which follows a formula, for given $t_0 \in \mathbb{T}$,

$$\mathfrak{F}(t) = \mathfrak{F}(t_0) + \int_{t_0}^{t} F(s) \Delta_g s \quad \text{for} \quad t \in \mathbb{T}.$$

Theorem 2.2. *Let $\{F_m\}_{m \in \mathbb{N}}$ be a sequence of rd-continuous functions from $[a, b]_{\mathbb{T}}$ into K_c^n.*

(i) *If $\{F_m\}$ uniformly converges to F on $[a, b]_{\mathbb{T}}$, then F is Δ_g-integrable and*

$$\int_a^b F(s)\Delta_g s = \lim_{m \to \infty} \int_a^b F_m(s)\Delta_g s.$$

(ii) *If $\{F_m\}$ converges to F at $t_0 \in [a, b]_{\mathbb{T}}$ and, for each m, F_m has continuous Δ_g-derivative $\Delta_g F_m$ such that $\{\Delta_g F_m\}$ converges uniformly to G, then F is Δ_g-differentiable and*

$$\Delta_g F(t) = G(t) \quad \text{for} \quad t \in [a, b]_{\mathbb{T}}.$$

Moreover, the sequence $\{F_m\}$ converges uniformly to F on $[a, b]_{\mathbb{T}}$.

Proof. (i) Let $\varepsilon > 0$. In virtue of the uniform convergence of $\{F_m\}$, there exists natural number m_0 such that

$$D[F_m(t), F(t)] < \frac{\varepsilon}{b - a}$$

for all $m > m_0$ and each $t \in [a, b]_{\mathbb{T}}$. Therefore,

$$D\left[\int_a^b F_m(s)\Delta_g s, \int_a^b F(s)\Delta_g s\right] \leq \int_a^b D[F_m(s), F(s)]\Delta s$$

$$\leq \frac{\varepsilon}{b - a}(b - a).$$

That is,

$$D\left[\int_a^b F_m(s)\Delta_g s, \int_a^b F(s)\Delta_g s\right] \leq \varepsilon.$$

Thus, we get the desired equality.

(ii) Since F_m is Δ_g-antiderivative of $\Delta_g F_m$, for each $t \in [a, b]_{\mathbb{T}}$, we have

$$F_m(t) = F_m(t_0) + \int_{t_0}^{t} \Delta_g F_m(s) \Delta_g s.$$

Letting m tends to ∞ in the above equation, in view of (i), we obtain

$$F(t) = F(t_0) + \int_{t_0}^{t} G(s) \Delta_g s.$$

Now, Theorem 2.1 shows

$$\Delta_g F'(t) = G(t) \quad \text{for} \quad t \in [a, b]_{\mathbb{T}}.$$

This completes the proof.   □

Chapter 3

Solvability of Set Dynamic Equations

3.1 Introduction

This chapter is mainly concerned with the existence of solutions to certain set dynamic equations on time scales, such as, the initial value problem, set functional equations with infinite delay, impulsive problem, and almost periodic solutions. Moreover, the phase space and (almost) periodicity on time scales are discussed.

The study of periodicity of solutions is one of the most interesting and important topics in the qualitative theory of differential equations and that of difference equations due to its mathematical interest and applications in different areas, such as in engineering, life sciences, information sciences and control theory, among others. However, the conditions to guarantee the periodicity are very demanding. For this reason, in the past decades, many authors have studied several extensions of the concept of periodicity and have established the existence of solutions with any such property for differential equations and their difference counterparts. We discuss almost periodicity and asymptotically almost periodicity properties. A careful investigation reveals that it is similar to the exploring some qualitative properties of almost periodic differential equations and their discrete analogy in the approaches. It is natural to propose the concept of almost periodic time scales and almost periodic functions on time scales to unify the study of the above-mentioned continuous and discrete problems and offer more general conclusions.

We also establish the phase space for the set dynamic equations with infinite delay which is an extension of the corresponding concept for single-valued dynamic problems on time scales. We systematically consider the existence of periodic solutions for a class of set dynamic equations with infinite delay on time scales, which generalize, and incorporate as special cases, some known results for set differential equations and for set difference equations, when the time scale is $\mathbb{R}$ and $\mathbb{Z}$, respectively. Moreover, for differential inclusions and difference inclusions when the variable under consideration is a single-valued mapping.

Several phenomena in physics, biology, and control theory during their evolutionary processes are subject to the action of short-time forces in the form of impulses. In most cases, the duration of the action of these forces is negligibly small, as a result of which one can assume that the forces act only at certain moments. The impulsive differential equations represent a mathematical model of such processes. In this chapter, a class of new nonlinear impulsive set dynamic equations is considered based on the new generalized derivative of set-valued functions developed on time scales. Some novel criteria are established for the existence and stability of solutions of such models. The approaches generalize, and incorporate as special cases, many known results for set (or fuzzy) differential equations and difference equations, when the time scale is $\mathbb{R}$ and $\mathbb{Z}$, respectively.

3.2 Initial Value Problem

We outline the steps of an alternative approach that leads to the study of a wide class of SDEs on time scales. We consider the following initial value problem of set integro-differential equations on time scales (SIDEs)

$$\begin{cases} \Delta_g U(t) = F\left(t, U(t), \int_0^t G(t, s) U(s) \Delta_g s\right), \\ U(0) = U_0 \in K_c^1, \end{cases} \tag{3.1}$$

where $F\colon J_{\mathbb{T}} \times K_c^1 \times K_c^1 \to K_c^1$, $G\colon J_{\mathbb{T}} \times J_{\mathbb{T}} \to \mathbb{R}^+$ with $J_{\mathbb{T}} := [0, a] \cap \mathbb{T}$, $a > 0$. We assume that $0, a \in \mathbb{T}$. We will apply monotone iterative technique to find the minimal and maximal solutions of (3.1). To construct the set monotone sequence, we first recall some notions, see Ahmad and Sivasundaram (2006) and Hong (2009).

Let P denote the subfamily of K_c^1 consisting of sets $A \in K_c^1$ with $A \subset \mathbb{R}^+$ and P^0 denote the subfamily of K_c^1 consisting of sets $A \in K_c^1$ with $A \subset \mathbb{R}^+ \setminus \{0\}$. Thus, P is a cone in K_c^1 and P^0 is the nonempty interior of the cone P. Now, we define the ordering in K_c^1.

Definition 3.1. For $A, B \in K_c^1$, we write $A \geq B$ or $B \leq A$ provided there exists $C \in K_c^1$ such that $C \subset P$ and $A = B + C$.

Definition 3.2. For $A, B \in K_c^1$, we write $A > B$ or $B < A$ provided there exists $C \in K_c^1$ such that $C \subset P^0$ and $A = B + C$.

Similarly, we can define the reverse ordering. The following theorem is concerning the order in K_c^1.

Theorem 3.1. *Assume that*

(h1) $F(\cdot, X(\cdot), Y(\cdot)) \in C_{\mathrm{rd}}[J_{\mathbb{T}}, K_c^1]$ *for each* $(X, Y) \in K_c^1 \times K_c^1$, *and* $G \colon J_{\mathbb{T}} \times J_{\mathbb{T}} \to \mathbb{R}^+$ *is continuous. In addition,*

$$m_0 := \max\{G(t, s) \colon t, s \in J_{\mathbb{T}}\} > 0.$$

(h2) $F(t, U, V)$ *is monotone nondecreasing in* $(U, V) \in K_c^1 \times K_c^1$ *for each* $t \in J_{\mathbb{T}}$, *that is, for fixed* $t \in J_{\mathbb{T}}$,

$$F(t, U_1, V_1) \geq F(t, U_2, V_2)$$

whenever $U_1 \geq U_2$ *and* $V_1 \geq V_2$.

(h3) *There exist* $V, W \in C_{\mathrm{rd}}^1[J_{\mathbb{T}}, K_c^1]$ *such that*

$$\Delta_g V(t) \leq F\left(t, V, \int_0^t G(t, s) V(s) \Delta_g s\right) \quad \text{for } t \in J_{\mathbb{T}},$$

and

$$\Delta_g W(t) \geq F\left(t, W, \int_0^t G(t, s) W(s) \Delta_g s\right) \quad \text{for } t \in J_{\mathbb{T}}.$$

(h4) *For* $X_1, X_2, Y_1, Y_2 \in K_c^1$, *with* $X_1 \geq X_2$, $Y_1 \geq Y_2$, $t, s \in J_{\mathbb{T}}$, *and some positive number* $L \geq m_0$, *we have*

$$F(t, X_1, Y_1) \leq F(t, X_2, Y_2) + L(X_1 - X_2) + L(Y_1 - Y_2).$$

Then, $V(t) \leq W(t)$ *for* $t \in J_{\mathbb{T}}$ *provided* $V(0) \leq W(0)$.

Proof. For $\varepsilon > 0$, we define

$$W_\varepsilon(t) = W(t) + \varepsilon e^{2Lt}$$

with L given in (h4). Then we observe that

$$V(0) \leq W(0) \leq W_\varepsilon(0).$$

Let $t_\delta \in J_{\mathbb{T}}$ be the supremum of all positive numbers δ such that $V(0) \leq W(0)$ implies

$$V(t) \leq W_\varepsilon(t) \quad \text{for all } t \in [0,\delta]_{\mathbb{T}}.$$

Evidently, $V(t_\delta) \leq W_\varepsilon(t_\delta)$ and $t_\delta \geq 0$. From Proposition 2.3(I), it follows that V and W are continuous. In view of the rd-continuity of V and W, it follows that $t_\delta > 0$ if 0 is a right-dense point. In addition, if t_δ is right-dense, then

$$\Delta_g W_\varepsilon(t_\delta) = \Delta_g W(t_\delta) + 2L\varepsilon e^{2Lt_\delta},$$

and if t_δ is right-scattered, then

$$\Delta_g W_\varepsilon(t_\delta) = \Delta_g W(t_\delta) + \varepsilon \frac{e^{2L\sigma(t_\delta)} - e^{2Lt_\delta}}{\mu(t_\delta)}.$$

First, we will show that $t_\delta = a$. On the contrary, suppose $\sigma(t_\delta) \leq a$. Our assumptions guarantee that

$$\Delta_g V(t_\delta) \leq F\left(t_\delta, V(t_\delta), \int_0^{t_\delta} G(t_\delta, s) V(s) \Delta_g s\right)$$

$$\leq F\left(t_\delta, W_\varepsilon(t_\delta), \int_0^{t_\delta} G(t_\delta, s) W_\varepsilon(s) \Delta_g s\right)$$

$$\leq F\left(t_\delta, W(t_\delta), \int_0^{t_\delta} G(t_\delta, s) W(s) \Delta_g s\right) + L(W_\varepsilon(t_\delta) - W(t_\delta))$$

$$+ L \int_0^{t_\delta} G(t_\delta, s)(W_\varepsilon(s) - W(s)) \Delta_g s$$

$$\leq F\left(t_\delta, W(t_\delta), \int_0^{t_\delta} G(t_\delta, s) W(s) \Delta_g s\right) + L\varepsilon e^{2Lt_\delta}$$

$$+ L \int_0^{t_\delta} m_0 \varepsilon e^{2Ls} \Delta s$$

$$\leq F\left(t_\delta, W(t_\delta), \int_0^{t_\delta} G(t_\delta, s)W(s)\Delta_g s\right) + L\varepsilon e^{2Lt_\delta}$$

$$+ L\frac{m_0\varepsilon}{2L}(e^{2Lt_\delta} - 1)$$

$$\leq F\left(t_\delta, W(t_\delta), \int_0^{t_\delta} G(t_\delta, s)W(s)\Delta_g s\right) + L\varepsilon e^{2Lt_\delta}$$

$$+ \frac{L\varepsilon}{2}(e^{2Lt_\delta} - 1)$$

$$\leq \Delta_g W(t_\delta) + 2L\varepsilon e^{2Lt_\delta}.$$

Further, it is easy to see

$$2Le^{2Lt_\delta} \leq \frac{e^{2L\sigma(t_\delta)} - e^{2Lt_\delta}}{\mu(t_\delta)}.$$

Hence, from the above inequality, we obtain

$$\Delta_g V(t_\delta) \leq \Delta_g W_\varepsilon(t_\delta). \tag{3.2}$$

If t_δ is a right-dense point, then Proposition 2.3(III), together with (3.2), implies that there exists $\eta > 0$ such that

$$W_\varepsilon(t_\delta) - V(t_\delta) \leq W_\varepsilon(t) - V(t) \quad \text{for } t_\delta < t < t_\delta + \eta.$$

This contradicts to the fact that $t_\delta > 0$ is the supremum. Hence, we get that $t_\delta = a$. Also, in view of the rd-continuity of V and W, we see that the inequality

$$V(t) < W_\varepsilon(t) \quad \text{holds for } t \in J_{\mathbb{T}}.$$

On the other hand, if t_δ is a right-scattered point, from (3.2) and Proposition 2.3(II), it follows that

$$\frac{V(\sigma(t_\delta)) - V(t_\delta)}{\mu(t_\delta)} \leq \frac{W_\varepsilon(\sigma(t_\delta)) - W_\varepsilon(t_\delta)}{\mu(t_\delta)},$$

i.e.,

$$W_\varepsilon(t_\delta) - V(t_\delta) \leq W_\varepsilon(\sigma(t_\delta)) - V(\sigma(t_\delta)).$$

Clearly, $t_\delta < \sigma(t_\delta) \leq a$ (otherwise, $t_\delta \geq a$ and the proof is completed). This shows that

$$W_\varepsilon(\sigma(t_\delta)) \geq V(\sigma(t_\delta))$$

which contradicts to the fact that t_δ is the supremum. Consequently, we obtain that

$$V(t) < W_\varepsilon(t) \quad \text{holds for all } t \in J_\mathbb{T}.$$

Thus, in any case, we have

$$V(t) < W_\varepsilon(t) \quad \text{for } t \in J_\mathbb{T}.$$

Now, taking the limit $\varepsilon \to 0$ yields the desired result. This completes the proof. $\qquad\square$

Corollary 3.1. *Let $V, W \in C_{\mathrm{rd}}^1[J_\mathbb{T}, K_c^1]$ such that*

$$\Delta_g V(t) \leq \Delta_g W(t) \quad \text{for all } t \in J_\mathbb{T}.$$

Then, $V(0) \leq W(0)$ implies that

$$V(t) \leq W(t) \quad \text{for all } t \in J_\mathbb{T}.$$

Definition 3.3. The functions $V, W \in C_{\mathrm{rd}}^1[J_\mathbb{T}, K_c^1]$ are said to be a lower solution and an upper solution of (3.1) respectively, provided

$$\begin{cases} \Delta_g V(t) \leq F\left(t, V(t), \displaystyle\int_0^t G(t,s)V(s)\Delta_g s\right), \\ V(0) \leq U_0 \in K_c^1, \end{cases} \tag{3.3}$$

and

$$\begin{cases} \Delta_g W(t) \geq F\left(t, W(t), \displaystyle\int_0^t G(t,s)W(s)\Delta_g s\right), \\ W(0) \geq U_0 \in K_c^1, \end{cases} \tag{3.4}$$

Definition 3.4. A set-valued function $U \colon J_\mathbb{T} \to K_c^1$ is said to be a solution of (3.1), provided it is Δ_g-differentiable and satisfies (3.1) on $J_\mathbb{T}$. The functions $\Phi \colon J_\mathbb{T} \to K_c^1$ and $\Psi \colon J_\mathbb{T} \to K_c^1$ are said to be the minimal solution and the maximal solution of (3.1), respectively, provided they both are solutions of (3.1) and

$$\Phi(t) \leq U(t) \leq \Psi(t)$$

for every solution U of (3.1) with

$$V(t) \leq U(t) \leq W(t) \quad \text{for } t \in J_\mathbb{T},$$

where V and W are lower solution and upper solution of (3.1) respectively, with $V(t) \leq W(t)$ for all $t \in J_\mathbb{T}$.

We are now in a position to state and prove the result concerning the existence of solutions of (3.1).

Theorem 3.2. *Assume that*

(s1) *(3.1) has lower solution V and upper solution W with $V, W \in C^1_{\mathrm{rd}}[J_{\mathbb{T}}, K^1_c]$ and $V(t) \leq W(t)$ for all $t \in J_{\mathbb{T}}$.*

(s2) *$F \colon J_{\mathbb{T}} \times K^1_c \times K^1_c \to K^1_c$ is continuous and maps bounded sets into bounded sets in K^1_c.*

(s3) *Hypotheses (h1), (h2), and (h4) hold.*

Then, there exist monotone sequences $\{V_n(t)\}$ and $\{W_n(t)\}$ in K^1_c such that

$$V_n(t) \to \Phi(t) \quad and \quad W_n(t) \to \Psi(t) \quad as\ n \to \infty \quad in\ K^1_c,$$

where Φ and Ψ are the minimal and maximal solutions of (3.1), respectively.

Proof. Let us construct the set integro-differential sequences by

$$
\begin{cases}
V_{n+1}(t) = U_0 + \displaystyle\int_0^t F\left(\tau, V_n(\tau), \int_0^\tau G(\tau, s) V_n(s) \Delta_g s\right) \Delta_g \tau \\[4mm]
W_{n+1}(t) = U_0 + \displaystyle\int_0^t F\left(\tau, W_n(\tau), \int_0^\tau G(\tau, s) W_n(s) \Delta_g s\right) \Delta_g \tau,
\end{cases}
$$

$$(3.5)$$

for $n \in \mathbb{N}_0$. Here, we stipulate $V_0(t) = V(t)$, $W_0(t) = W(t)$, $t \in J_{\mathbb{T}}$. Note that, according to Theorem 2.1(i), the set-valued functions V_{n+1} and W_{n+1} are all well-defined. Moreover, in view of Theorem 2.1(i), we have that

$$
\begin{cases}
\Delta_g V_{n+1}(t) = F\left(t, V_n(t), \displaystyle\int_0^t G(t, s) V_n(s) \Delta_g s\right), \\[4mm]
V_{n+1}(0) = U_0 \in K^1_c,
\end{cases}
$$

$$(3.6)$$

and

$$
\begin{cases}
\Delta_g W_{n+1}(t) = F\left(t, W_n(t), \displaystyle\int_0^t G(t, s) W_n(s) \Delta_g s\right), \\[4mm]
W_{n+1}(0) = U_0 \in K^1_c,
\end{cases}
$$

$$(3.7)$$

Now, we prove that

$$V_0(t) \leq V_1(t) \leq W_1(t) \leq W_0(t)$$

for $t \in J_{\mathbb{T}}$. Since V_0 is lower solution of (3.1), V_0 satisfies (3.3) and $V_0(0) \leq U_0$. In virtue of Corollary 3.1, inequality (3.3), and Eq. (3.6), we have

$$V_0(t) \leq V_1(t) \quad \text{for all } t \in J_{\mathbb{T}}.$$

Following the method of this proof, it is easy to see that

$$W_1(t) \leq W_0(t) \quad \text{on } J_{\mathbb{T}}.$$

Substituting $n = 0$ in (3.6) and (3.7), by means of the monotone property of F, we obtain

$$\Delta_g V_1(t) = F\left(t, V_0(t), \int_0^t G(t,s)V_0(s)\Delta_g s\right)$$

$$\leq F\left(t, W_0(t), \int_0^t G(t,s)W_0(s)\Delta_g s\right)$$

$$= \Delta_g W_1(t), \quad t \in J_{\mathbb{T}}.$$

Since $V_1(0) = W_1(0) = U_0$, from Corollary 3.1, we have

$$V_1(t) \leq W_1(t) \quad \text{for all } t \in J_{\mathbb{T}}.$$

Consequently, we arrive at the conclusion that

$$V_0(t) \leq V_1(t) \leq W_1(t) \leq W_0(t) \quad \text{for all } t \in J_{\mathbb{T}}.$$

Next, we inductively assume

$$V_{j-1}(t) \leq V_j(t) \leq W_j(t) \leq W_{j-1}(t) \quad \text{for } t \in J_{\mathbb{T}} \text{ and } j > 1.$$

We shall prove that

$$V_j(t) \leq V_{j+1}(t) \leq W_{j+1}(t) \leq W_j(t) \quad \text{for all } t \in J_{\mathbb{T}}.$$

To end this, substituting $n = j$ in (3.6) and (3.7), by means of the monotone property of F, we obtain

$$\Delta_g V_j(t) = F\left(t, V_{j-1}(t), \int_0^t G(t,s)V_{j-1}(s)\Delta_g s\right)$$

$$\leq F\left(t, V_j(t), \int_0^t G(t,s)V_j(s)\Delta_g s\right)$$

$$= \Delta_g V_{j+1}(t) \quad \text{for } t \in J_{\mathbb{T}},$$

and

$$\Delta_g W_j(t) = F\left(t, W_{j-1}(t), \int_0^t G(t,s) W_{j-1}(s) \Delta_g s\right)$$

$$\geq F\left(t, W_j(t), \int_0^t G(t,s) W_j(s) \Delta_g s\right)$$

$$= \Delta_g W_{j+1}(t) \quad \text{for } t \in J_{\mathbb{T}}.$$

From this, combining $V_j(0) = V_{j+1}(0) = W_j(0) = W_{j+1}(0) = U_0$, and by virtue of Corollary 3.1, we get

$$V_j(t) \leq V_{j+1}(t) \ \text{ and } \ W_{j+1}(t) \leq W_j(t) \quad \text{for } t \in J_{\mathbb{T}}.$$

Again, by means of the monotone property of F and our inductive assumption, we have

$$\Delta_g V_{j+1}(t) = F\left(t, V_j(t), \int_0^t G(t,s) V_j(s) \Delta_g s\right)$$

$$\leq F\left(t, W_j(t), \int_0^t G(t,s) W_j(s) \Delta_g s\right).$$

That is,

$$\Delta_g V_{j+1}(t) \leq \Delta_g W_{j+1}(t) \quad \text{for } t \in J_{\mathbb{T}}.$$

Thus, Corollary 3.1 guarantees that

$$V_{j+1}(t) \leq W_{j+1}(t) \quad \text{for all } t \in J_{\mathbb{T}}.$$

Hence, by induction, we get

$$V_0(t) \leq V_1(t) \leq \cdots \leq V_n(t) \leq \cdots$$
$$\leq W_n(t) \leq \cdots \leq W_1(t) \leq W_0(t) \quad \text{for } t \in J_{\mathbb{T}}. \tag{3.8}$$

Next, from (3.5), Theorem 2.1(i), and Proposition 2.3(I), it follows that V_n and W_n are continuous for $n \in \mathbb{N}_0$. Since $J_{\mathbb{T}}$ is compact, the sequences $\{V_n(t)\}$ and $\{W_n(t)\}$ are uniformly bounded on $J_{\mathbb{T}}$. Also, in virtue of the condition (s2), it is easy to see that they both are equicontinuous on $J_{\mathbb{T}}$. Now, from (3.8) and the standard application of Theorem A.1, it follows that the sequence $\{V_n\}$ converges uniformly to Φ and $\{W_n\}$ converges uniformly to Ψ on $J_{\mathbb{T}}$. Clearly, $\Phi, \Psi \in \mathcal{C}_{rd}^1(J_{\mathbb{T}}, K_c^1)$. On the other hand, we have

$$D\left[\int_0^t G(t,s)V_n(s)\Delta_g s, \int_0^t G(t,s)\Phi(s)\Delta_g s\right]$$

$$\leq \int_0^t G(t,s)D[V_n(s), \Phi(s)]\Delta s$$

$$\leq am_0 \max_{s \in J_{\mathbb{T}}} D[V_n(s), \Phi(s)].$$

This shows that the limit

$$\lim_{n \to \infty} \int_0^t G(t,s)V_n(s)\Delta_g s = \int_0^t G(t,s)\Phi(s)\Delta_g s$$

holds uniformly on $J_{\mathbb{T}}$. Now, condition (s2) guarantees that F is uniformly continuous on $J_{\mathbb{T}} \times \{V_n(t)\} \times \{W_n(t)\}$ and the sequence

$$\left\{F\left(t, V_n(t), \int_0^t G(t,s)V_n(s)\Delta_g s\right)\right\}$$

converges uniformly to

$$F\left(t, \Phi(t), \int_0^t G(t,s)\Phi(s)\Delta_g s\right)$$

on $J_{\mathbb{T}}$. Using this and passing to the limit $n \to \infty$ in the first equation of (3.5), we find that Φ is a solution of (3.1). Similarly, we can see that Ψ is also a solution of (3.1). Moreover, by means of (3.8) we easily get

$$V_0(t) \leq \Phi(t) \leq \Psi(t) \leq W_0(t) \quad \text{for } t \in J_{\mathbb{T}}.$$

Finally, we show that Φ and Ψ are the minimal and maximal solutions of (3.1), respectively. To end this, we assume that U is any solution of (3.1) satisfying

$$V_0(t) \leq U(t) \leq W_0(t) \quad \text{for every } t \in J_{\mathbb{T}}.$$

Thus, we have to prove that

$$V_0(t) \leq \Phi(t) \leq U(t) \leq \Psi(t) \leq W_0(t) \quad \text{for } t \in J_{\mathbb{T}}. \tag{3.9}$$

First, we prove

$$V_1(t) \leq U(t) \leq W_1(t) \quad \text{for each } t \in J_{\mathbb{T}}.$$

Now, using (3.9) together with the monotone property of F, and (3.6) and (3.7), we find that

$$\Delta_g U(t) \geq F\left(t, V_0(t), \int_0^t G(t,s)V_0(s)\Delta_g s\right)$$
$$= \Delta_g V_1(t)$$

and $U(0) = U_0$. This, in virtue of Corollary 3.1, implies $V_1(t) \leq U(t)$ for each $t \in J_{\mathbb{T}}$. With the analogous method, we can get

$$U(t) \leq W_1(t) \quad \text{for each } t \in J_{\mathbb{T}}.$$

Moreover, the induction immediately implies that

$$V_n(t) \leq U(t) \leq W_n(t)$$

for $n \in \mathbb{N}_0$ and each $t \in J_{\mathbb{T}}$. Taking the limit $n \to \infty$ in the above inequality, (3.9) is established and this completes the proof. □

3.3 Phase Space and Periodicity

In this section, a phase space is built for set functional dynamic equations with infinite delay on time scales and sufficient criteria are established for the existence of periodic solutions of such equations. The results proved in this section generalize, and incorporate as special cases, some known results for set differential equations and for set difference equations, when the time scale is $\mathbb{R}$ and $\mathbb{Z}$, respectively.

3.3.1 *Periodic properties*

We first formulate some basic results for periodic functions on periodic time scales. Let $\nu \colon \mathbb{T} \to \mathbb{R}$ be a strictly increasing function such that $\widetilde{\mathbb{T}} = \nu(\mathbb{T})$ is also a time scale. By $\widetilde{\sigma}$, we denote the jump operator on $\widetilde{\mathbb{T}}$ and by $\widetilde{\Delta}_g$, we denote the Δ_g-derivative for functions on $\widetilde{\mathbb{T}}$. Then, we see that $\nu \circ \sigma = \widetilde{\sigma} \circ \nu$. The following results are necessary to discuss the periodic properties.

Theorem 3.3 (Chain rule). *Let $\nu\colon \mathbb{T} \to \mathbb{R}$ be a strictly increasing function such that $\widetilde{\mathbb{T}} = \nu(\mathbb{T})$ is also a time scale. Let $W\colon \widetilde{\mathbb{T}} \to K_c^1$. Suppose $\nu^\Delta(t)$ and $\widetilde{\Delta}_g(W \circ \nu)(t)$ exist for $t \in \mathbb{T}$. Then,*

$$\Delta_g(W \circ \nu) = (\widetilde{\Delta}_g W \circ \nu)\nu^\Delta.$$

Proof. For given $\varepsilon \in (0,1)$, let

$$\varepsilon^* = \frac{\varepsilon}{\left[1 + |\nu^\Delta(t)| + \|\widetilde{\Delta}_g W(\nu(t))\|\right]}.$$

Then, according to the hypothesis, there exist neighborhoods $U_{\mathbb{T}}$ of t and $V_{\mathbb{T}}$ of $\nu(t)$ such that

$$|\nu(\sigma(t)) - \nu(s) - (\sigma(t) - s)\nu^\Delta(t)| \le \varepsilon^*|\sigma(t) - s| \quad \text{for all } s \in U_{\mathbb{T}},$$

and

$$D[W(\nu(t+h)) - W(\widetilde{\sigma}(\nu(t))), (\nu(t+h) - \widetilde{\sigma}(\nu(t)))\widetilde{\Delta}_g W(\nu(t))]$$
$$\le \varepsilon^*|\nu(t+h) - \widetilde{\sigma}(\nu(t))|$$

for all $\nu(t+h) \in U_{\mathbb{T}}$. Let $h \in U_{\mathbb{T}}$ be such that $\nu(t+h) \in V_{\mathbb{T}}$ and $|h| > 0$. Then,

$$D[W(\nu(t+h)) - W(\nu(\sigma(t))), (\widetilde{\Delta}_g W(\nu(t))\nu^\Delta(t))(t+h-\sigma(t)))]$$

$$= D[W(\nu(t+h)) - W(\widetilde{\sigma}(\nu(t))) + (\nu(t+h) - \widetilde{\sigma}(\nu(t)))\widetilde{\Delta}_H W(\nu(t)),$$
$$\quad (\nu(t+h) - \widetilde{\sigma}(\nu(t)))\widetilde{\Delta}_g W(\nu(t))$$
$$\quad + (t+h-\sigma(t))\nu^\Delta(t)\widetilde{\Delta}_g W(\nu(t))]$$

$$\le D[W(\nu(t+h)) - W(\widetilde{\sigma}(\nu(t))), (\nu(t+h) - \widetilde{\sigma}(\nu(t)))\widetilde{\Delta}_g W(\nu(t))]$$
$$\quad + D[(\nu(t+h) - \widetilde{\sigma}(\nu(t)))\widetilde{\Delta}_g W(\nu(t)),$$
$$\qquad (t+h-\sigma(t))\nu^\Delta(t)\widetilde{\Delta}_g W(\nu(t))]$$

$$\le \varepsilon^*|\nu(t+h) - \widetilde{\sigma}(\nu(t))| + \varepsilon^*|\mu(t) - h|\|\widetilde{\Delta}_g W(\nu(t))\|$$

$$\le \varepsilon^*|\nu(t+h) - \widetilde{\sigma}(\nu(t)) - (t+h-\sigma(t))\nu^\Delta(t)|$$
$$\quad + |(h-\mu(t))\nu^\Delta(t)| + \varepsilon^*|\mu(t) - h|\|\widetilde{\Delta}_g W(\nu(t))\|$$

$$\le \varepsilon^*\left[\varepsilon^*|h-\mu(t)| + h-\mu(t)|\nu^\Delta(t)| + |h-\mu(t)|\|\widetilde{\Delta}_g W(\nu(t))\|\right]$$

$$\le \varepsilon^*\left[1 + \|\nu^\Delta(t)\| + \|\widetilde{\Delta}_g W(\nu(t))\|\right]|h-\sigma(t)|$$

$$= \varepsilon|h - \mu(t)|.$$

In the view of Definition 2.5, we obtain

$$\Delta_g(W \circ \nu)(t) = (\widetilde{\Delta}_g W \circ \nu)(t) \, \nu^{\Delta}(t) \quad \text{for} \quad t \in \mathbb{T},$$

which yields the desired equality. This completes the proof. □

Theorem 3.4 (Substitution). *Assume that $F \colon \mathbb{T} \to K_c^1$ is an rd-continuous set-valued function and $\nu \colon \mathbb{T} \to \mathbb{R}$ is delta differentiable with rd-continuous derivative. Then, for $a, b \in \mathbb{T}$,*

$$\int_a^b F(t)\nu^{\Delta}(t)\Delta_g t = \int_{\nu(a)}^{\nu(b)} (F \circ \nu^{-1})(s)\widetilde{\Delta}_g s.$$

Proof. Since $F\nu^{\Delta}$ is an rd-continuous function, by Remark 2.3, it possesses an antiderivative $\mathfrak{F}$, i.e., $\Delta_g \mathfrak{F} = F\nu^{\Delta}$. Now, Theorem 2.1 implies

$$\int_a^b F(t)\nu^{\Delta}(t)\Delta_g t = \int_a^b \Delta_g\mathfrak{F}(t)\Delta_g t$$

$$= \mathfrak{F}(b) - \mathfrak{F}(a)$$

$$= (\mathfrak{F} \circ \nu^{-1})(\nu(b)) - (\mathfrak{F} \circ \nu^{-1})(\nu(a))$$

$$= \int_{\nu(a)}^{\nu(b)} \widetilde{\Delta}_g(\mathfrak{F} \circ \nu^{-1})(s)\widetilde{\Delta}_g s.$$

Moreover, by means of Theorems 1.4 and 3.3, we have

$$\int_{\nu(a)}^{\nu(b)} \widetilde{\Delta}_g(\mathfrak{F} \circ \nu^{-1})(s)\widetilde{\Delta}_g s$$

$$= \int_{\nu(a)}^{\nu(b)} (\Delta_g\mathfrak{F} \circ \nu^{-1})(s)(\nu^{-1})^{\widetilde{\Delta}}\widetilde{\Delta}_g s$$

$$= \int_{\nu(a)}^{\nu(b)} ((F\nu^{\Delta}) \circ \nu^{-1})(s)(\nu^{-1})^{\widetilde{\Delta}}\widetilde{\Delta}_g s$$

$$= \int_{\nu(a)}^{\nu(b)} (F \circ \nu^{-1})(s)[(\nu^{\Delta} \circ \nu^{-1})(\nu^{-1})^{\widetilde{\Delta}}](s)\widetilde{\Delta}_g s$$

$$= \int_{\nu(a)}^{\nu(b)} (F \circ \nu^{-1})(s)\widetilde{\Delta}_g s.$$

This shows that the desired equality holds. □

Definition 3.5. Let $\omega > 0$. A time scale $\mathbb{T}$ is called ω-periodic if $t + \omega \in \mathbb{T}$ for any $t \in \mathbb{T}$. If $\mathbb{T} \neq \mathbb{R}$, the smallest $\omega > 0$ is called the period of the time scale $\mathbb{T}$. Also, by $\mathbb{T} + \omega \subset \mathbb{T}$, we mean that the set $\{t + \omega : t \in \mathbb{T}\}$ is contained in $\mathbb{T}$.

Definition 3.6. Let $\mathbb{T} \neq \mathbb{R}$ be a ω-periodic time scale. A function $p \colon \mathbb{T} \to K_c^1$ is said to be ω-periodic on $\mathbb{T}$ provided $p(t + \omega) = p(t)$ for each $t \in \mathbb{T}$.

If $\mathbb{T} = \mathbb{R}$, then p is said to be ω-periodic provided $\omega > 0$ is smallest such that $p(t + \omega) = p(t)$ for each $t \in \mathbb{T}$.

Theorem 3.5. *Suppose that* $\mathbb{T}$ *is* ω-*periodic time scale and* $p \in \mathcal{C}_{\mathrm{rd}}(\mathbb{T}, K_c^1)$ *is* ω-*periodic regressive function, and* $a, b \in \mathbb{T}$. *Then,*

(i) $\sigma(t + \omega) = \sigma(t) + \omega$, $\rho(t + \omega) = \rho(t) + \omega$, *and* $\mu(t + \omega) = \mu(t)$.

(ii) $e_p(t + \omega, s + \omega) = e_p(t, s)$ *and* $e_p(t + \omega, t)$ *is independent of* $t \in \mathbb{T}$.

(iii) $\displaystyle\int_{a+\omega}^{b+\omega} p(t) \Delta_g t = \int_a^b p(t) \Delta_g t.$

Proof. (i) The first two identity follow easily from the fact $\mathbb{T}$ is ω-periodic time scale. Next,

$$\mu(t + \omega) = \sigma(t + \omega) - (t + \omega)$$
$$= \sigma(t) + \omega - t - \omega$$
$$= \sigma(t) - t$$
$$= \mu(t).$$

(ii) Since $p \in \mathcal{C}_{\mathrm{rd}}(\mathbb{T}, K_c^1)$, using Definition 1.3, we have

$$e_p(t, s) = \exp\left\{ \int_s^t \xi_\mu(p(\tau)) \Delta_g \tau \right\}$$
$$= \exp\left\{ \int_{s+\omega}^{t+\omega} \xi_\mu(p(\tau)) \Delta_g \tau \right\}$$
$$= e_p(t + \omega, s + \omega).$$

Also, let $E(t) := e_p(t + \omega, t)$ for $t \in \mathbb{T}$ and $t_0 \in \mathbb{T}$. Then,

$$E(t) = e_p(t + \omega, t_0) e_p(t_0, t)$$
$$= e_p(t, t_0 - \omega) e_p(t_0, t).$$

Taking the Δ_g-derivative, we get

$$\Delta_g E(t) = \Delta_g(e_p(t, t_0 - w))e_p(t_0, t) + e_p^\sigma(t, t_0 - w)\Delta_g e_p(t_0, t)$$

$$= pe_p(t, t_0 - w)e_p(t_0, t) + (1 + \mu p(t))e_p(t, t_0 - w)(\ominus p)e_p(t_0, t)$$

$$= pE(t) - \frac{p}{1 + \mu p(t)}(1 + \mu p(t))E(t)$$

$$= 0.$$

This yields that $E(t)$ does indeed not depend on $t \in \mathbb{T}$. Consequently, $e_p(t + w, t)$ is independent of $t \in \mathbb{T}$.

(iii) We set $\nu(t) = t + w$. Then, ν is strictly increasing such that $\nu^\Delta(t) = 1$. Also, $\nu(\mathbb{T}) = \mathbb{T} + w =: \widetilde{\mathbb{T}}$. Since p is w-periodic on $\mathbb{T}$, in virtue of Theorem 3.3, we obtain

$$\int_a^b p(t)\Delta_g t = \int_a^b p(t)\nu^\Delta(t)\Delta_g t$$

$$= \int_{\nu(a)}^{\nu(b)} (p \circ \nu^{-1})(s)\widetilde{\Delta}_g s$$

$$= \int_{a+w}^{b+w} p(s - w)\Delta_g s$$

$$= \int_{a+w}^{a+w} p(s)\Delta_g s.$$

The proof is complete. $\square$

Theorem 3.6. *Let $\mathbb{T}$ be w-periodic time scale. Suppose that F: $\mathbb{T} \times \mathbb{T} \to K_c^1$ is continuous at (t, t), where $t \in \mathbb{T}$. Also, suppose that $\Delta_g^t F(t, \cdot)$ is rd-continuous on $[a, \sigma(t)]_{\mathbb{T}}$ and, for each $\varepsilon > 0$, there exists a neighborhood $U_{\mathbb{T}}$ of t, independent of $\tau \in [a, \sigma(t)]_{\mathbb{T}}$, such that*

$$D[F(t + h, \tau) - F(\sigma(t), \tau), \Delta_H^t F(t, \tau)(h - \mu(t))] \le \varepsilon|h - \mu(t)|$$

for all $t + h \in U_{\mathbb{T}}$, where $\Delta_g^t F$ denotes the Δ_g-derivative of F with respect to first variable t. Then,

$$G(t) = \int_t^{t+w} F(t, \tau)\Delta_g \tau \tag{3.10}$$

implies

$$\Delta_g G(t) = \int_t^{t+\omega} \Delta_g^t F(t,\tau)\Delta_g\tau + F(\sigma(t), t+\omega) - F(\sigma(t), t).$$

Proof. Take $a \in \mathbb{T}$ with $a > t$, from Proposition 2.5(i3), we have

$$G(t) = \int_t^a F(t,\tau)\Delta_g\tau + \int_a^{t+\omega} F(t,\tau)\Delta_g\tau.$$

Further, by Theorem 2.1(iii), we have

$$\Delta_g \left[\int_a^t F(t,\tau)\Delta_g\tau \right] = \int_a^t \Delta_g^t F(t,\tau)\Delta_g\tau + F(\sigma(t), t); \quad (3.11)$$

$$\Delta_g \left[\int_t^b F(t,\tau)\Delta_g\tau \right] = \int_t^b \Delta_g^t F(t,\tau)\Delta_g\tau - F(\sigma(t), t). \quad (3.12)$$

Therefore, it is sufficient to prove that the assertions hold, for $b, c \in \mathbb{T}$ with $b < t < c$. In fact, if (3.11) and (3.12) are true for $b, c \in \mathbb{T}$ with $b < t < c$, then, from (3.10), we have

$$G(t) = \int_t^{c+\omega} F(t,\tau)\Delta_g\tau + \int_{c+\omega}^{t+\omega} F(t,\tau)\Delta_g\tau$$

$$= \int_t^{c+\omega} F(t,\tau)\Delta_g\tau + \int_c^t F(t,\tau+\omega)\Delta_g\tau.$$

In what follows, by Theorem 2.1(iii), we arrive at

$$\Delta_g G(t) = \int_t^{c+\omega} \Delta_g^t F(t,\tau)\Delta_g\tau - F(\sigma(t), t)$$

$$+ \int_c^t \Delta_g^t F(t,\tau+\omega)\Delta_g\tau + F(\sigma(t), t+\omega)$$

$$= \int_t^{t+\omega} \Delta_g^t F(t,\tau)\Delta_g\tau + F(\sigma(t), t+\omega) - F(\sigma(t), t).$$

Thus, the desired equality achieved and this completes the proof. $\square$

3.3.2 The phase spaces

It is well known that the development of functional differential equations with infinite delay primarily depends on the choice of phase space. The phase space for initial functions also plays a very important role in the study of functional dynamic equations with infinite delay on time scales. A phase space for functional dynamic equations with infinite delay on time scales is established in Bi *et al.* (2008).

Suppose that $\inf \mathbb{T} = -\infty$ and $t_1, t_2 \in \mathbb{T}$ are such that $t_1 + t_2 \in \mathbb{T}$. Let $\mathbb{R}^+ = [0, +\infty)$, $\mathbb{R}^- = (-\infty, 0]$, $\mathbb{T}^- = \mathbb{T} \cap \mathbb{R}^-$, $\mathbb{T}^+ = \mathbb{T} \cap \mathbb{R}^+$, and

$$\Xi := \left\{ \xi \in \mathcal{C}_{\mathrm{rd}}(\mathbb{T}^-, \mathbb{R}^+) \colon \xi(s) > 0 \text{ for all } s \in \mathbb{T}^- \text{and} \int_{-\infty}^0 \xi(s)\Delta s = 1 \right\}.$$

The phase space C_ξ with $\xi \in \Xi$ is defined by

$$C_\xi := \left\{ \varphi \in \mathcal{C}_{\mathrm{rd}}(\mathbb{T}^-, \mathbb{R}) \colon \int_{-\infty}^0 \xi(s)|\varphi|^{[s,0]}\Delta s < \infty \right\}$$

and is endowed with the norm

$$|\varphi|_\xi = \int_{-\infty}^0 \xi(s)|\varphi|^{[s,0]}\Delta s,$$

where

$$|\varphi|^{[s,0]} = \sup\{|\varphi(t)| \colon s \le t \le 0\}.$$

In this section, we establish a phase space for set functional dynamic equations with infinite delay on time scales. To this end, we first define the distance $D^{I_\mathbb{T}}$ on $\mathcal{C}_{\mathrm{rd}}(I_\mathbb{T}, K_c^1)$ by

$$D^{I_\mathbb{T}}[F, G] = \sup_{t \in I_\mathbb{T}} D[F(t), G(t)]$$

for any $F, G \in \mathcal{C}_{\mathrm{rd}}(I_\mathbb{T}, K_c^1)$, where $I_\mathbb{T} \subset \mathbb{T}$ is a bounded closed subset. By RS_F, we denote the collection of all rd-continuous selectors of F.

Theorem 3.7. *For $F \in \mathcal{C}_{\mathrm{rd}}(I_\mathbb{T}, K_c^1)$ and $f \in RS_F$, there exists positive number $\eta = \eta(F) < 1$ such that for each $G \in \mathcal{C}_{\mathrm{rd}}(I_\mathbb{T}, K_c^1)$, we can find $g \in RS_G$ satisfying*

$$\eta\|f - g\| \le D^{I_\mathbb{T}}[F, G],$$

where $\|f - g\| = \sup\{|f(t) - g(t)| \colon t \in I_\mathbb{T}\}$.

Proof. For given $F, G \in \mathcal{C}_{\mathrm{rd}}(I_{\mathbb{T}}, K_c^1)$, it is clear that there exists $\mu \in (0, 1)$ such that

$$\mu D^{I_{\mathbb{T}}}[RS_F, RS_G] \leq D^{I_{\mathbb{T}}}[F, G].$$

On the other hand, for any given $t_0 \in I_{\mathbb{T}}$ and $f \in RS_F$, from Lemma 2.1, there exists $g_{t_0} \in RS_G$ such that

$$\mu|f(t_0) - g_{t_0}(t_0)| \leq D[RS_F(t_0), RS_G(t_0)], \tag{3.13}$$

where $RS_W(t) = \{w(t) \colon w \in RS_W\}$. In virtue of continuity of the metric D, one can find $\delta_0 > 0$ such that (3.13) holds on the interval $(t_0 - \delta_0, t_0 + \delta_0)_{\mathbb{T}}$. Since $I_{\mathbb{T}}$ is bounded and closed, there exist finitely many intervals, say, $I_i =: (t_i - \delta_i, t_i + \delta_i)_{\mathbb{T}}$, $i = 1, 2, \cdots, n$, $n \in \mathbb{N}$, $\delta_i > 0$, such that

$$I_{\mathbb{T}} \subset \bigcup_{i=1}^{n} I_i$$

and

$$\mu|f(t) - g_i(t)| \leq D[RS_F(t), RS_G(t)] \quad \text{for all } t \in I_i$$

with $g_i \in RS_G$, $(i = 1, 2, \ldots, n)$. Let

$$\widehat{g}(t) = \begin{cases} g_1(t) & \text{for } t \in I_1 \\ g_2(t) & \text{for } t \in I_2 \\ \vdots \\ g_n(t) & \text{for } t \in I_n. \end{cases}$$

It is clear that $\widehat{g} \in RS_G$ and for any $t \in I_{\mathbb{T}}$, we have $t \in I_i$ for some i such that

$$\mu|f(t) - \widehat{g}(t)| = \mu|f(t) - g_i(t)|$$
$$\leq D[RS_F(t), RS_G(t)].$$

Now, from the arbitrariness of $t \in \mathbb{T}$, it follows that

$$\mu\|f - \widehat{g}\| \leq D^{I_{\mathbb{T}}}[RS_F, RS_G].$$

Let $\eta = \mu^2$. Then, $0 < \eta < 1$ and

$$\eta\|f - \widehat{g}\| \leq D^{I_{\mathbb{T}}}[F, G].$$

This completes the proof. $\qquad\qquad\qquad\qquad\qquad\qquad\qquad\square$

Let $\mathbb{B}$ be the collection of bounded nonempty subsets of $\mathbb{T}$. Then, $\gamma \colon \mathbb{B} \to \mathbb{R}^+$ defined by

$$\gamma(I_{\mathbb{T}}) = \inf \left\{ \eta(F) \in (0,1) \colon F \in \mathcal{C}_{\mathrm{rd}}(I_{\mathbb{T}}, K_c^1) \text{ admits } f \in RS_F \right.$$

$$\text{such that for any } G \in \mathcal{C}_{\mathrm{rd}}(I_{\mathbb{T}}, K_c^1) \text{ there exists}$$

$$\left. g \in RS_G \text{ satisfying } \eta(F)\|f - g\| \leq D^{I_{\mathbb{T}}}[F, G] \right\}$$

is called the measure of continuity on $I_{\mathbb{T}}$. Obviously, Theorem 3.7 guarantees that γ exists and $0 \leq \gamma < 1$.

Theorem 3.8. *Let $I_{\mathbb{T}} \in \mathbb{B}$ be a closed subset of $\mathbb{T}$. If $\gamma(I_{\mathbb{T}}) > 0$, then the space $\mathcal{C}_{\mathrm{rd}}(I_{\mathbb{T}}, K_c^1)$ is a complete metric space when endowed with the metric $D^{I_{\mathbb{T}}}$.*

Proof. Let $\{F_n\} \subset \mathcal{C}_{\mathrm{rd}}(\mathbb{T}^-, K_c^1)$ be a Cauchy sequence. Then, for given $\varepsilon > 0$, there exists a nature number N_ε such that for $m, n \geq N_\varepsilon$, we have

$$D^{I_{\mathbb{T}}}[F_m, F_n] < \varepsilon.$$

Now, for $m, n \geq N_\varepsilon$ and $t \in I_{\mathbb{T}}$, we have

$$D[F_m(t), F_n(t)] < \varepsilon.$$

Thus, $\{F_n(t)\} \subset K_c^1$ is a Cauchy sequence and hence, from the completeness of K_c^1, converges to, say, $F(t)$. Now, we can assume that, for any $n \geq N_\varepsilon$ and $t \in I_{\mathbb{T}}$,

$$D[F_n(t), F(t)] < \frac{\varepsilon}{3}. \tag{3.14}$$

Let $\varepsilon > 0$ be given and $t_0 \in I_{\mathbb{T}}$ be right-dense. Since F_{N_ε} is rd-continuous, there exists a neighborhood Ω of t_0 such that for any $t \in \Omega$, we have

$$D[F_{N_\varepsilon}(t_0), F_{N_\varepsilon}(t)] < \frac{\varepsilon}{3}.$$

From this, we infer

$$D[F(t_0), F(t)] \leq D(F(t_0), F_{N_\varepsilon}(t_0)) + D[F_{N_\varepsilon}(t_0), F_{N_\varepsilon}(t)]$$
$$+ D[F_{N_\varepsilon}(t), F(t)]$$
$$< \varepsilon.$$

It follows that F is continuous at the right-dense point t_0.

If t_0 is left-dense, there exists $A \in K_c^1$ such that, for some $\delta > 0$,

$$D[F_{N_\varepsilon}(t), A] < \frac{\varepsilon}{3} \quad \text{for all } t \in (t_0 - \delta, t_0)_\mathbb{T}.$$

Now, for $t \in (t_0 - \delta, t_c)_\mathbb{T}$, we have

$$D[F(t), A] \leq D[F(t), F_{N_\varepsilon}(t)] + D[F_{N_\varepsilon}(t), A]$$
$$< \varepsilon.$$

Therefore, F is continuous at each right-dense point and its left-side limit exists at each left-dense point. Since, for fixed $n \in \mathbb{N}$,

$$D[F_n(t), F(t)]$$

is continuous with respect to $t \in I_\mathbb{T}$ and $I_\mathbb{T}$ is compact, we can assure that, for any $n \geq N_\varepsilon$,

$$D^{I_\mathbb{T}}[F_n, F] < \frac{\varepsilon}{3}.$$

Let $\varepsilon^* = \frac{\gamma(I)\varepsilon}{3}$. Then, in virtue of Theorem 3.7, for given $n \in \mathbb{N}$, we can find $\eta = \eta(F_n) \in (0, 1)$ such that, for any $m \in \mathbb{N}$ and $f_n \in RS_{F_n}$, there exists $f_m \in RS_{F_m}$ such that

$$\eta|f_m(t) - f_n(t)| \leq D^{I_\mathbb{T}}[F_m, F_n] < \varepsilon^*.$$

This, combining with $\gamma(I_\mathbb{T}) \leq \eta(F_n)$, implies that

$$|f_m(t) - f_n(t)| < \frac{\varepsilon}{3} \quad \text{for all } t \in I_\mathbb{T}. \tag{3.15}$$

Hence, $\{f_n(t)\}$ is Cauchy sequence and moreover, converges to, say, $f(t)$ for each $t \in I_\mathbb{T}$. Thus,

$$|f_n(t) - f(t)| \leq \frac{\varepsilon}{3} \quad \text{for } n \geq N_\varepsilon. \tag{3.16}$$

We assert that $f(t) \in F(t)$. In fact, from (3.14) and (3.16), we obtain

$$d(f(t), F(t)) \leq |f(t) - f_n(t)| + d(f_n(t), F(t))$$
$$\leq |f(t) - f_n(t)| + D[F_n(t), F(t)]$$
$$\leq \frac{\varepsilon}{3} + \frac{\varepsilon}{3}.$$

This yields,

$$d(f(t), F(t)) < \varepsilon.$$

Now, since $F(t)$ is a compact set, by the arbitrariness of ε, we obtain

$$d(f(t), F(t)) = 0,$$

which implies $f(t) \in F(t)$. Next, we show that $f \in RS_F$. To see this, we suppose that $\varepsilon > 0$ is given and $t^* \in I_{\mathbb{T}}$ is right-dense. Since f_{N_ε} is rd-continuous, there exists a neighborhood Ω of t^* such that for any $t \in \Omega$, we have

$$|f_{N_\varepsilon}(t^*) - f_{N_\varepsilon}(t)| < \frac{\varepsilon}{3}.$$

Combining this with (3.16), we have

$$|f(t^*) - f(t)| \le |f(t^*) - f_{N_\varepsilon}(t^*)| + |f_{N_\varepsilon}(t^*) - f_{N_\varepsilon}(t)|$$
$$+ |f_{N_\varepsilon}(t) - f(t)|$$
$$< \varepsilon.$$

This implies that f is continuous at right-dense point t^*. If t^* is left-dense, again, applying the rd-continuity of f_{N_ε}, we can conclude that there exist $\delta > 0$ and $a \in \mathbb{R}$ such that

$$|f_{N_\varepsilon}(t) - a| < \frac{\varepsilon}{3} \quad \text{for all } t \in (t^* - \delta, t^*)_{\mathbb{T}}.$$

This, combining with (3.16), follows that

$$|f(t) - a| \le |f(t) - f_{N_\varepsilon}(t)| + |f_{N_\varepsilon}(t) - a|$$
$$< \varepsilon.$$

Thus, f has finite limit a at left-dense point t^*. Consequently, $f \in RS_F$. Finally, we prove $F \in \mathcal{C}_{\mathrm{rd}}(I_{\mathbb{T}}, K_c^1)$. Let n be so large that (3.16) is satisfied. Then, for any fixed $G \in \mathcal{C}_{\mathrm{rd}}(I_{\mathbb{T}}, K_c^1)$, there exists $g_n \in RS_G$ such that, for all $t \in I_{\mathbb{T}}$,

$$\gamma(I)|f_n(t) - g_n(t)| \le D^{I_{\mathbb{T}}}[F_n, G].$$

Since $G(t)$ is compact, we can find a subsequence of $\{g_n(t)\}$ (without loss of generality, we may take $\{g_n(t)\}$ itself) such that

$$\lim_{n \to \infty} g_n(t) = g(t)$$

uniformly for $t \in I_{\mathbb{T}}$. It is easy to see that $g \in RS_G$. Let n tends to infinity. Then, from the above inequality, it follows that

$$\gamma(I_{\mathbb{T}})|f(t) - g(t)| \leq D^{I_{\mathbb{T}}}[F, G].$$

This shows that $F \in \mathcal{C}_{\mathrm{rd}}(I_{\mathbb{T}}, K_c^1)$ which completes the proof. $\qquad \square$

For any $T > 0$, let $I_T = [-T, 0]_{\mathbb{T}^-}$. By $\mathscr{C}(I_T, K_c^1)$, we mean the space $\mathcal{C}_{\mathrm{rd}}(I_T, K_c^1)$ to satisfy that the measure of continuity on I_T is positive, i.e., $\gamma(T) =: \gamma(I_T) > 0$. Let us define

$$\mathscr{C}_\xi := \big\{ \Phi \in \mathcal{C}_{\mathrm{rd}}(\mathbb{T}^-, K_c^1) \colon D_\xi[\Phi, \{0\}]\Delta s < \infty \text{ and}$$

$$\Phi|_{I_T} \in \mathscr{C}(I_T, K_c^1) \text{ for all } T > 0 \big\},$$

where $\xi \in \Xi$ and

$$D_\xi[F, G] := \int_{-\infty}^0 \xi(s) D^{[s,0]}[F, G]\Delta s.$$

It is clear that $D_\xi[F, G] < \infty$ if $F, G \in \mathscr{C}_\xi$ and $D_\xi[\cdot, \cdot]$ is a distance defined on $\mathscr{C}_\xi$.

Theorem 3.9. *Let $\xi \in \Xi$. Then the following results are true.*

(i) *For any $\varepsilon > 0$ and $T > 0$, there exists $\delta = \delta(\varepsilon, T) > 0$ such that, for any $\Phi_1, \Phi_2 \in \mathscr{C}_\xi$, the inequality*

$$D_\xi[\Phi_1, \Phi_2] \leq \delta$$

implies

$$D^{I_T}[\Phi_1, \Phi_2] \leq \varepsilon.$$

(ii) *Suppose $\{F_n\} \subset \mathcal{C}_{\mathrm{rd}}(\mathbb{T}^-, K_c^1)$ satisfies*

$$D^{I_s}[F_n, \theta] \leq \varphi(s) \quad \text{for } s > 0 \quad \text{and} \quad n \in \mathbb{N},$$

where $\varphi \colon \mathbb{T} \to \mathbb{R}$ is positive function such that

$$\int_{-\infty}^0 \xi(s)\varphi(s)\Delta s < \infty.$$

Then, $\lim_{n \to \infty} D_\xi(F_n, F) = 0$ if and only if $\lim_{n \to \infty} D^{I_{n_0}}(F_n, F) = 0$ for any $n_0 \in \mathbb{N}$.

(iii) *The space $(\mathscr{C}_\xi, D_\xi)$ is a complete metric space.*

Proof. (i) On the contrary, suppose there exist $\varepsilon_0 > 0$ and $T_0 > 0$ such that, for any $\delta > 0$, there exist $\Phi_1, \Phi_2 \in \mathscr{C}_\xi$ such that

$$D_\xi[\Phi_1, \Phi_2] \leq \delta$$

but

$$D^{[-T_0, 0]}[\Phi_1, \Phi_2] > \varepsilon_0.$$

Now, choosing $\delta = \frac{\varepsilon_0}{2} \int_{-\infty}^{-T_0} \xi(s) \Delta s$, we have

$$\delta \geq D_\xi[\Phi_1, \Phi_2]$$

$$= \int_{-\infty}^{0} \xi(s) D^{[s, 0]}[\Phi_1, \Phi_2] \Delta s$$

$$\geq \int_{-\infty}^{-T_0} \xi(s) D^{[s, 0]}[\Phi_1, \Phi_2] \Delta s$$

$$\geq \int_{-\infty}^{-T_0} \xi(s) D^{[-T_0, 0]}[\Phi_1, \Phi_2] \Delta s$$

$$> \varepsilon_0 \int_{-\infty}^{-T_0} \xi(s) \Delta s$$

$$= 2\delta.$$

This is a contradiction. Hence, (i) is true.

(ii) The necessity of (ii) is easily seen by means of (i). For the sufficiency, let $\{F_n\}$ and φ be given as in the assumptions of (ii) and suppose

$$\lim_{n \to \infty} D^{I_{n_0}}[F_n, F] = 0 \quad \text{for all } n_0 \in \mathbb{N}.$$

Thus, we can assume that

$$D^{I_s}[F_n, \theta] \leq \varphi(s) \quad \text{for any } s > 0.$$

For any $\varepsilon > 0$, since

$$\int_{-\infty}^{0} \xi(s) \varphi(s) \Delta s < \infty,$$

there exists $n_0 \in \mathbb{N}$ such that

$$\int_{-\infty}^{n_0} \xi(s)\varphi(s)\Delta s < \varepsilon.$$

In virtue of the standard Lebesgue dominated convergence theorem, we have

$$\lim_{n \to \infty} D_\xi[F_n, F]$$

$$= \lim_{n \to \infty} \int_{-\infty}^{0} \xi(s) D^{[s,0]}[F_n, F]\Delta s$$

$$= \lim_{n \to \infty} \int_{-\infty}^{-n_0} \xi(s) D^{[s,0]}[F_n, F]\Delta s + \lim_{n \to \infty} \int_{-n_0}^{0} \xi(s) D^{[s,0]}[F_n, F]\Delta s$$

$$\leq 2 \lim_{n \to \infty} \int_{-\infty}^{-n_0} \xi(s)\varphi(s)\Delta s + \lim_{n \to \infty} \int_{-n_0}^{0} \xi(s) D^{[-n_0,0]}[F_n, F]\Delta s$$

$$< 2\varepsilon + \int_{-n_0}^{0} \xi(s) \lim_{n \to \infty} D^{[-n_0,0]}[F_n, F]\Delta s$$

$$= 2\varepsilon.$$

From the arbitrariness of ε, it follows that

$$\lim_{n \to \infty} D_\xi[F_n, F] = 0.$$

(iii) Let $\{F_n\}$ be Cauchy sequence in $\mathscr{C}_\xi$. Then, for any given $s > 0$, from this and (i), it follows that there exist $\delta > 0$ and $n_0 \in \mathbb{N}$ such that, for any $n \geq n_0$,

$$D_\xi[F_n, F_{n_0}] \leq \delta \quad \text{and} \quad D^{I_s}[F_n, F_{n_0}] < 1.$$

Thus,

$$D^{I_s}[F_n, \theta] \leq D^{I_s}[F_n, F_{n_0}] + D^{I_s}[F_{n_0}, \theta]$$

$$\leq 1 + D^{I_s}[F_{n_0}, \theta].$$

Let $\varphi(s) = 1 + D^{I_s}[F_{n_0}, \theta]$. Then, $\varphi(s) > 0$ and

$$\int_{-\infty}^{0} \xi(s)\varphi(s)\Delta s = \int_{-\infty}^{0} \xi(s)\Delta s + \int_{-\infty}^{0} \xi(s) D^{[s,0]}[F_{n_0}, \theta]\Delta s$$

$$= 1 + D_\xi[F_{n_0}, \theta]$$

$$< \infty.$$

In virtue of (ii), we get

$$\lim_{m,n \to \infty} D^{I_T}[F_m, F_n] = 0 \quad \text{for } T > 0,$$

whenever

$$\lim_{m,n \to \infty} D_\xi[F_m, F_n] = 0.$$

Therefore, $\{F_n\}$ is a Cauchy sequence in $\mathscr{C}(I_T, K_c^1)$. Now, Theorem 3.8 guarantees that there exists $F \in \mathscr{C}(I_T, K_c^1)$ such that

$$D^{I_T}[F_n, F] \to 0 \quad \text{as } n \to \infty.$$

Moreover, in virtue of (ii), we have

$$D_\xi[F_n, F] \to 0 \quad \text{as } n \to \infty.$$

In addition, for any $\varepsilon > 0$, there exists $N \in \mathbb{N}$ such that

$$D_\xi[F_n, F] < \frac{\varepsilon}{2} \quad \text{and} \quad D^{I_T}[F_n, F] < \frac{\varepsilon}{2} \quad \text{for all } n \geq N.$$

In view of this, we infer that

$$D_\xi[F, \theta] = \int_{-\infty}^{0} \xi(s) D^{[s,0]}[F, \theta] \Delta s$$

$$\leq \int_{-\infty}^{0} \xi(s) D^{[s,0]}[F, F_N] \Delta s + \int_{-\infty}^{0} \xi(s) D^{[s,0]}[F_N, \theta] \Delta s$$

$$< \frac{\varepsilon}{2} + D_\xi[F_N, \theta]$$

$$< \infty.$$

Now, it remains to show that $F \in \mathcal{C}_{\text{rd}}(\mathbb{T}^-, K_c^1)$. Since $F_n \in \mathcal{C}_{\text{rd}}(\mathbb{T}^-, K_c^1)$, in view of Theorem 3.7, there exists $f_n \in RS_{F_n}$ such that for each n,

$$\gamma(I_T)|f_m(t) - f_n(t)| \leq D^{I_T}[F_m, F_n]$$

for any $T > 0$ and $t \in I_T$. Also, from the fact

$$\lim_{m,n \to \infty} D^{I_T}[F_m, F_n] = 0,$$

it follows that $\{f_n\}$ is a Cauchy sequence on I_T. Therefore, there exists an rd-continuous function f such that f_n uniformly converges to f on I_T. Now, from the arbitrariness of T, it follows that f is rd-continuous in $\mathbb{T}^-$. As an analogy of the proof of Theorem 3.8, we have $f \in RS_F$. The remainder is to observe $F_n|_{I_T} \in \mathscr{C}(I_T, K_c^1)$ and

$$D^{I_T}[F_n|_{I_T}, F|_{I_T}] \to 0 \quad \text{as } n \to \infty$$

for all $T > 0$. The completeness of $\mathscr{C}(I_T, K_c^1)$ yields $F|_{I_T} \in \mathscr{C}(I_T, K_c^1)$. Consequently, $F \in \mathscr{C}_\xi$. This completes the proof. $\square$

Theorem 3.10. *Suppose that $\xi \in \Xi$, $\Phi \in \mathscr{C}_\xi$, and $U_t(\chi) = U(t+\chi)$ for $\chi \in \mathbb{T}^-$.*

(i) *Let $T > 0$. Assume that $U \colon (-\infty, T)_{\mathbb{T}} \to K_c^1$ with $U|_{[0,T]_{\mathbb{T}}} \in \mathscr{C}([0,T]_{\mathbb{T}}, K_c^1)$ satisfies $U_0 = \Phi$. Then, for any $t \in [0,T]_{\mathbb{T}}$,*

$$U_t \in \mathscr{C}_\xi \quad and \quad V(t) = U_t \in \mathscr{C}([0,T]_{\mathbb{T}}, K_c^1).$$

(ii) *There exists a nonnegative constant M such that*

$$\|\Phi(0)\| \le M D_\xi[\Phi, \theta].$$

Proof. (i) From Proposition 2.5, we obtain, for any $t \in [0,T]_{\mathbb{T}}$, that

$$D_\xi[U_t, \theta] = \int_{-\infty}^0 \xi(s) D^{[s,0]}[U_t, \theta] \Delta s$$

$$= \int_{-\infty}^{-t} \xi(s) D^{[s,0]}[U_t, \theta] \Delta s + \int_{-t}^0 \xi(s) D^{[s,0]}[U_t, \theta] \Delta s$$

$$\le \int_{-\infty}^{-t} \xi(s) \max\left(D^{[s+t,0]}[U, \theta], D^{[-t,0]}[U_t, \theta] \right) \Delta s$$

$$\quad + \int_{-t}^0 \xi(s) D^{[s,0]}[U_t, \theta] \Delta s$$

$$\le \int_{-\infty}^0 \xi(s) D^{[s+t,0]}[U, \theta] \Delta s + 2 \int_{-\infty}^0 \xi(s) D^{[0,t]}[U|_{[0,T]_{\mathbb{T}}}, \theta] \Delta s$$

$$\le \int_{-\infty}^0 \xi(s) D^{[s,0]}[\Phi, \theta] \Delta s + 2 \int_{-\infty}^0 \xi(s) D^{[0,T]}[U|_{[0,T]_{\mathbb{T}}}, \theta] \Delta s$$

$$= D_\xi[\Phi, \theta] + 2 D^{[0,T]}[U|_{[0,T]_{\mathbb{T}}}, \theta] \times \int_0^{-\infty} \xi(s) \Delta s$$

$$< \infty.$$

In addition, it is clear that $U_t \in \mathcal{C}_{\mathrm{rd}}(\mathbb{T}^-, K_c^1)$ and

$$U|_{[-T+t,t]_\mathbb{T}} \in \mathscr{C}([-T+t,t]_\mathbb{T}, K_c^1),$$

which is equivalent to

$$U_t|_{I_T} \in \mathscr{C}(I_T, K_c^1).$$

Hence,

$$U_t \in \mathscr{C}_\xi \quad \text{for} \quad t \in [0,T]_\mathbb{T}.$$

Next, we show that V is rd-continuous on $[0,T]_\mathbb{T}$. Let $t_0 \in [0,T]_\mathbb{T}$ be a right-dense point. Since $U_{t_0} \in \mathscr{C}_\xi$, in view of Lemma 2.2, we see that U is bounded on $[0,T]_\mathbb{T}$, i.e., there exists $L > 0$ such that

$$D^{[0,T]}[U,\theta] \leq L.$$

On the other hand, for any $\varepsilon > 0$, there exists $M = M(t_0, \varepsilon) > 0$ such that

$$\int_{-\infty}^{-M} \xi(s) D^{[s,0]}[U_{t_0}, \theta] \Delta s < \frac{\varepsilon}{4} \quad \text{and} \quad \int_{-\infty}^{-M} \xi(s) \Delta s < \frac{\varepsilon}{4L}.$$

Moreover,

$$U_{t_0} \in \mathscr{C}_\xi \quad \text{implies that } U_{t_0} \in \mathcal{C}_{\mathrm{rd}}(\mathbb{T}^-, K_c^1).$$

Let $r \in [-M, 0]$. If r is right-dense, then we can choose sufficiently small $\delta_1 > 0$ such that, for any $t \in (t_0 - \delta_1, t_0 + \delta_1)_\mathbb{T}$,

$$D(U_t(r), U_{t_0}(r)) = D(U_t(r), U_t(r - t + t_0))$$

$$< \frac{\varepsilon}{4}.$$

If r is right-scattered and left-dense, then U_{t_0} has finite left-sided limit at r, say A. Now, by definition of rd-continuity, there exists $\delta_2 > 0$ such that

$$D[U_{t_0}(s), A] < \frac{\varepsilon}{8} \quad \text{for all } s \in (r - \delta_2, r + \delta_2)_\mathbb{T}.$$

Let $|t - t_0| < \delta_2$. Then, $r + t_0 - t \in (r - \delta_2, r + \delta_2)_\mathbb{T}$ and

$$D[U_t(r), U_{t_0}(r)] = D[U_t(r), U_t(r + t_0 - t)]$$

$$\leq D[U_t(r), A] + D[A, U_t(r + t_0 - t)]$$

$$< \frac{\varepsilon}{8} + \frac{\varepsilon}{8}.$$

That is,

$$D[U_t(r), U_{t_0}(r)] < \frac{\varepsilon}{4}.$$

Finally, if r is right-scattered and left-scattered, then it is clear that there exists $\delta_3 > 0$ such that $s = r$, whenever $s \in (r - \delta_3, r + \delta_3)_{\mathbb{T}}$. Thus, $r + t_0 - t = r$ when $|t - t_0| < \delta_3$. Hence, for $|t - t_0| < \delta_3$,

$$D[U_t(r), U_{t_0}(r)] = D[U_t(r), U_t(r + t_0 - t)] = 0.$$

Consequently,

$$D[U_t(r), U_{t_0}(r)] < \frac{\varepsilon}{4}.$$

As a summary, for each $r \in [-M, 0]_{\mathbb{T}}$, we have that

$$D^{[-M,0]}[U_t, U_{t_0}] < \frac{\varepsilon}{4}$$

if $t \in (t_0 - \delta, t_0 + \delta)_{\mathbb{T}}$ with $\delta := \min\{\delta_1, \delta_2, \delta_3\}$. Now, we have

$$
\begin{aligned}
D_\xi[U_t, U_{t_0}] &= \int_{-\infty}^{0} \xi(s) D^{[s,0]}[U_t, U_{t_0}] \Delta s \\
&= \int_{-\infty}^{-M} \xi(s) D^{[s,0]}[U_t, U_{t_0}] \Delta s + \int_{-M}^{0} \xi(s) D^{[s,0]}[U_t, U_{t_0}] \Delta s \\
&\leq \int_{-\infty}^{-M} \xi(s) \left(D^{[s,0]}[U_t, \theta] + D^{[s,0]}[U_{t_0}, \theta] \right) \Delta s \\
&\quad + \int_{-M}^{0} \xi(s) D^{[s,0]}(U_t, U_{t_0}) \Delta s \\
&\leq \int_{-\infty}^{-M} \xi(s) \left(\max\{D^{[t_0,t]}[U, \theta] + D^{[s,0]}[U_{t_0}, \theta]\} \right. \\
&\quad \left. + D^{[s,0]}[U_{t_0}, \theta] \right) \Delta s + \int_{-M}^{0} \xi(s) D^{[s,0]}[U_t, U_{t_0}] \Delta s \\
&\leq \int_{-\infty}^{-M} \xi(s) \left(L + 2D^{[s,0]}[U_{t_0}, \theta] \right) \Delta s + D^{[-M,0]}[U_t, U_{t_0}] \\
&\leq L\frac{\varepsilon}{4L} + 2\frac{\varepsilon}{4} + \frac{\varepsilon}{4}.
\end{aligned}
$$

That is,

$$D_\xi[U_t, U_{t_0}] < \varepsilon \quad \text{if } t \in (t_0 - \delta, t_0 + \delta)_\mathbb{T}.$$

This implies that U_t is continuous at right-dense point t_0 with respect to $t \in [0, T]_\mathbb{T}$. If t_0 is left-dense point, then we can similarly find that U_t has a finite limit at t_0 with respect to $t \in [0, T]_\mathbb{T}$. Since $U_t \in \mathscr{C}_\xi$, we infer that, for each fixed $t \in [0, T]_\mathbb{T}$, there exists $u_t \in RS_{U_t}$. Since u_t is rd-continuous on $\mathbb{T}^-$, by the similar argument as the above, we can see that $v \in RS_V$ with $v(t) = u_t$ for $t \in [0, T]_\mathbb{T}$. Thus, V is rd-continuous on $[0, T]_\mathbb{T}$, i.e., $V \in \mathcal{C}_{\mathrm{rd}}([0, T]_\mathbb{T}, K_c^1)$.

We are now in a position to prove that $V \in \mathscr{C}([0, T]_\mathbb{T}, K_c^1)$. For this, it is sufficient to check that, for fixed $v \in RS_V$ and any $G \in \mathscr{C}([0, T]_\mathbb{T}, K_c^1)$, there exists $g \in RS_G$ such that

$$\gamma([0, T]_\mathbb{T})\|g - v\| \le D^{[0,T]}[V, G].$$

Let

$$\bar{G}(t) = \begin{cases} G(t) & \text{for } t \in (0, T]_\mathbb{T} \\ \Phi(t) & \text{for } t \in \mathbb{T}^-. \end{cases}$$

Then, $\bar{G}_t \in \mathscr{C}_\xi$ and $\bar{G}_0 = \Phi$, in addition $G(t) = \bar{G}_t(0)$ and $V(t) = U_t(0)$. Also, it has been shown that, for fixed $t \in [0, T]_\mathbb{T}$, if $v(t) = u_t \in RS_{U_t}$, then there exists $\bar{g}_t \in RS_{\bar{G}_t}$ such that

$$\gamma(M) \max_{s \in I_M} |\bar{g}_t(s) - u_t(s)| \le D^{I_M}[\bar{G}_t, U_t]$$

for all $M > 0$. Since $\gamma(M)$ is decreasing, we have

$$0 < \gamma = \lim_{M \to 0} \gamma(M) \le 1$$

and

$$\gamma|\bar{g}_t(0) - u_t(0)| \le D[\bar{G}_t(0), U_t(0)]$$
$$= D[G(t), V(t)].$$

Let $g(t) = \bar{g}_t$, $v(t) = u_t$ for $t \in [0, T]_\mathbb{T}$. Then,

$$\gamma|g(t) - v(t)| = |\bar{g}_t(0) - u_t(0)|$$
$$\le D[G(t), V(t)]$$
$$\le D^{[0,T]}[G, V].$$

By the arbitrariness of $t \in [0, T]_{\mathbb{T}}$, we have

$$\gamma \max_{t \in [0,T]_{\mathbb{T}}} |g(t) - v(t)| \leq D^{[0,T]}[V, G].$$

This completes the proof of (i).

(ii) The assertion is obvious, since we have

$$\|\Phi(0)\| = \int_{-\infty}^{0} \xi(s) \|\Phi(0)\| \Delta s$$

$$\leq \int_{-\infty}^{0} \xi(s) D^{[s,0]}(\Phi, \theta) \Delta s$$

$$= D_{\xi}(\Phi, \theta).$$

This completes the proof. $\qquad\qquad\qquad\qquad\qquad\qquad\qquad$ □

Remark 3.1. If the functions under consideration are single-valued, then the $\mathscr{C}_{\xi}$ space is the phase space C_{ξ} ($\xi \in \Xi$) established by Bi *et al.* (2008) for functional dynamic equation with infinite delay on $\mathbb{T}$. Moreover, if $\mathbb{T} = \mathbb{R}$, then $\mathscr{C}_{\xi}$ is the phase space for ordinary functional differential equations with infinite delay.

Let $\mathscr{C}_{\omega} = \{U \in C_{\mathrm{rd}}(\mathbb{T}, K_c^1) \colon U(t + \omega) = U(t)\}$ and $\Omega = [0, \omega]_{\mathbb{T}}$.

Remark 3.2. From Lemma 2.2, it follows that U is bounded, if $U \in \mathscr{C}_{\omega}$.

Lemma 3.1. *Suppose that $\{U^n\} \subset \mathscr{C}_{\omega}$, $n \in \mathbb{N}$, and $U \in \mathscr{C}_{\omega}$ are such that*

$$D^{\Omega}[U^n, U] \to 0 \quad as \ n \to \infty.$$

Then, $\{U_t^n\}$ converges uniformly to $U_t \in \mathscr{C}_{\xi}$ with respect to t.

Proof. From the hypothesis, for given $\varepsilon > 0$, there exists $N \in \mathbb{N}$ such that

$$D^{\Omega}[U^n, U] < \varepsilon \quad \text{for } n \geq N.$$

Thus, for $n \geq N$, we have

$$D_{\xi}[U_t^n, U_t] = \int_{-\infty}^{0} \xi(s) D^{[s,0]}[U_t^n, U_t] \Delta s$$

$$= \int_{-\infty}^{0} \xi(s) D^{[s+t,t]}[U^n, U] \Delta s$$

$$\leq D^{\Omega}[U^n, U] \int_{-\infty}^{0} \xi(s) \Delta s$$

$$< \varepsilon.$$

This completes the proof. □

3.3.3 *Periodic solutions*

In this section, we investigate the existence of periodic solutions for the set functional dynamic equation with infinite delay on time scales (SFDEID) of the form

$$\Delta_g U(t) = -a(t, U(t)) U(\sigma(t)) + F(t, U_t), \quad t \in \mathbb{T}. \qquad (3.17)$$

Here, $s \in \mathbb{T}^-$ and $t \in \mathbb{T}$ are such that $s + t \in \mathbb{T}$ and $U_t(s) = U(t+s)$, and $U \in \mathcal{C}_{\mathrm{rd}}^1(\mathbb{T}, K_c^1)$, $a \in C_{\mathrm{rd}}(\mathbb{T} \times K_c^1), \mathbb{R})$ and $F \colon \mathbb{T} \times K_c^1 \to K_c^1$ is a set-valued function.

We will employ the monotone method to present some interesting observations. For any $U, V \in K_c^1$, the ordering $U \geq V$ (or $V \leq U$) is according to Definition 3.1. In this section, we assume that $0 \in \mathbb{T}$. For $\xi \in \Xi$, we introduce the following notation,

$$\mathscr{C}_{\mathbb{T}} = \{ U \in \mathscr{C}_\omega \colon U_t \in \mathscr{C}_\xi \text{ for any } t \in \mathbb{T} \}.$$

Thus,

$$U|_{\mathbb{T}^-} = U_0 \in \mathscr{C}_\xi \quad \text{provided } U \in \mathscr{C}_{\mathbb{T}}.$$

For any $\Phi, \Psi \in \mathscr{C}_{\mathbb{T}}$, we write $\Phi \leq \Psi$ whenever $\Phi(t) \leq \Psi(t)$ for each $t \in \mathbb{T}$.

Lemma 3.2. *The function $D^{I_{\mathbb{T}}}[\cdot, \theta]$ is increasing with respect to K_c^1. That is, for any $U, V \in K_c^1$, the inequality $U \leq V$ implies*

$$D^{I_{\mathbb{T}}}[U, \theta] \leq D^{I_{\mathbb{T}}}[V, \theta].$$

Proof. Let $V = U + W$ with $U, V, W \in K_c^1$. For $u \in U$ and $w \in W$, denote

$$v(t) = u(t) + w(t) \quad \text{for } t \in I_{\mathbb{T}}.$$

Then, $v \in V$ and $v(t) \geq u(t) \geq 0$ for $t \in I_{\mathbb{T}}$. This yields that

$$u(t) \leq D[V(t), \theta]$$
$$\leq D^{I_{\mathbb{T}}}[V, \theta].$$

From the arbitrariness of t and u, it follows that

$$D^{I_{\mathbb{T}}}[U, \theta] \leq D^{I_{\mathbb{T}}}[V, \theta].$$

This completes the proof. □

Definition 3.7. An rd-continuous set-valued function $U \colon \mathbb{T} \to K_c^1$ is said to be a solution of SFDEID (3.17) provided U is Δ_g-differentiable, and satisfies SFDEID (3.17).

Definition 3.8. Two functions $V, W \in \mathscr{C}_{\mathbb{T}}$ are said to be a lower solution and an upper solution of SFDEID (3.17), respectively, provided V and W satisfy

$$D^{\Omega}[V, \theta] < \infty, \quad D^{\Omega}[W, \theta] < \infty, \quad V(0) \leq W(0),$$

and

$$\Delta_g V(t) \leq -a(t, V(t))V(\sigma(t)) + F(t, V_t),$$

and

$$\Delta_g W(t) \geq -a(t, W(t))W(\sigma(t)) + F(t, W_t), \quad t \in \mathbb{T}.$$

Definition 3.9. Two functions $\Phi \colon \mathbb{T} \to K_c^1$ and $\Psi \colon \mathbb{T} \to K_c^1$ are said to be the minimal solution and the maximal solution of (3.17), respectively, provided they both are solutions of SFDEID (3.17) and

$$\Phi(t) \leq U(t) \leq \Psi(t)$$

for every solution U of SFDEID (3.17) with

$$V(t) \leq U(t) \leq W(t) \quad \text{for } t \in \mathbb{T},$$

where V and W are lower solution and upper solution of SFDEID (3.17), respectively, with $V(t) \leq W(t)$ for all $t \in \mathbb{T}$.

The following lemma is essentially important.

Theorem 3.11. *Let* $b: \mathbb{T} \to \mathbb{R}$ *be an* ω-*periodic, regressive, and rd-continuous, and* $P \in \mathscr{C}_\omega$. *Then, SDE*

$$\Delta_g U(t) = -b(t)U(\sigma(t)) + P(t) \tag{3.18}$$

has a unique ω-*periodic solution* U *given by*

$$U(t) = \frac{1}{k_b} \int_t^{t+\omega} P(s)e_b(s,t)\Delta_g s \quad for \ t \in \mathbb{T}, \tag{3.19}$$

where $k_b = e_b(t+\omega, t) - 1$.

Proof. We first prove that U defined by (3.19) is an ω-periodic solution of SDE (3.18). Theorem 3.5 shows that k_b is independent of t and

$$U(t+\omega) = \frac{1}{k_b} \int_{t+\omega}^{t+2\omega} P(s)e_b(s, t+\omega)\Delta_g s$$

$$= \frac{1}{k_b} \int_t^{t+\omega} P(s+\omega)e_b(s+\omega, t+\omega)\Delta_g s$$

$$= \frac{1}{k_b} \int_t^{t+\omega} P(s)e_b(s, t)\Delta_g s$$

$$= U(t).$$

This implies that U is ω-periodic. Next, in view of Theorem 3.6, we have

$$\Delta_g U(t) = \frac{1}{k_b} \int_t^{t+\omega} P(s)(\ominus b)(t)e_b(s, t)\Delta_g s$$

$$+ \frac{1}{k_b}\left(P(t+\omega)e_b(t+\omega, \sigma(t)) - P(t)e_b(t, \sigma(t))\right)$$

$$= \frac{-b(t)}{k_b} \int_t^{t+\omega} P(s)e_b(s, \sigma(t))\Delta_g s$$

$$+ \frac{1}{k_b}\left(P(t)e_b(t+\omega, \sigma(t)) - P(t)e_b(t, \sigma(t))\right).$$

Also, from Theorem 3.6, it follows that

$$k_b b(t) U(\sigma(t))$$

$$= b(t) \int_{\sigma(t)}^{\sigma(t)+\omega} P(s) e_b(s, \sigma(t)) \Delta_g s$$

$$= b(t) \left(\int_{\sigma(t)}^{t} P(s) e_b(s, \sigma(t)) \Delta_g s + \int_{t}^{t+\omega} P(s) e_b(s, \sigma(t)) \Delta_g s \right.$$

$$\left. + \int_{t+\omega}^{\sigma(t)+\omega} P(s) e_b(s, \sigma(t)) \Delta_g s \right)$$

$$= b(t) \left(\int_{t}^{t+\omega} P(s) e_b(s, \sigma(t)) \Delta_g s - \int_{t}^{\sigma(t)} P(s) e_b(s, \sigma(t)) \Delta_g s \right.$$

$$\left. + \int_{t+\omega}^{\sigma(t+\omega)} P(s) e_b(s, \sigma(t)) \Delta_g s \right)$$

$$= b(t) \left(\int_{t}^{t+\omega} P(s) e_b(s, \sigma(t)) \Delta_g s - \mu(t) P(t) e_b(t, \sigma(t)) \right.$$

$$\left. + \mu(t+\omega) P(t) e_b(t+\omega, \sigma(t)) \right).$$

Hence,

$$k_b(\Delta_g U(t) + b(t) U(\sigma(t)))$$

$$= P(t) e_b(t+\omega, \sigma(t)) - P(t) e_b(t, \sigma(t))$$

$$\quad + b(t) \left(-\mu(t) P(t) e_b(t, \sigma(t)) + \mu(t+\omega) P(t) e_b(t+\omega, \sigma(t)) \right)$$

$$= P(t) \left((1 + \mu(t) b(t)) e_b(t+\omega, \sigma(t)) - (1 + \mu(t) b(t)) e_b(t, \sigma(t)) \right)$$

$$= P(t)(e_b(t+\omega, t) - e_b(t, t))$$

$$= k_b P(t).$$

This shows that U solves SDE (3.18).

Conversely, assume that U is ω-periodic and solves (3.18). Let $t_0 \in \mathbb{T}$ be arbitrarily given. Then, from Theorem 1.6, it follows that

$$\Delta_g[U(t) e_b(t, t_0)] = \Delta_g U(t) e_b(t, t_0) + U(\sigma(t)) e_b^\Delta(t, t_0)$$

$$= e_b(t, t_0)(\Delta_g U(t) + b(t) U(\sigma(t)))$$

$$= e_b(t, t_0) P(t).$$

That is,

$$\Delta_g[U(t)e_b(t, t_0)] = e_b(t, t_0)P(t).$$

Now, integrating both sides of the above equation from t to $t + \omega$ and using the property of Δ_g- integral, we have

$$\int_t^{t+\omega} e_b(s, t_0)P(t)\Delta_g s = U(t + \omega)e_b(t + \omega, t_0) - U(t)e_b(t, t_0)$$

$$= U(t)(e_b(t + \omega, t_0) - e_b(t, t_0))$$

$$= U(t)e_b(t, t_0)k_b.$$

The arbitrariness of $t_0 \in \mathbb{T}$ implies that U satisfies (3.19). This completes the proof.   □

Theorem 3.12. *Consider the SDE* (3.17). *Assume that V and W are the upper and lower solutions of* (3.17), *respectively. Suppose*

(h1) *for $\Phi, \Psi \in \mathscr{C}_\mathbb{T}$ with $\Phi \leq \Psi$, there exists some positive number L such that the inequalities*

$$0 \leq F(t, \Psi_t) - F(t, \Phi_t) - (a(t, \Psi(t))\Psi(\sigma(t)) - a(t, \Phi(t))\Phi(\sigma(t)))$$

$$\leq L(\Psi(t) - \Phi(t))$$

hold for $t \in \mathbb{T}$.

Then, $V \leq W$ on $\mathbb{T}$.

Proof. For any $\varepsilon > 0$, we define

$$W_\varepsilon(t) = W(t) + \varepsilon e^{Lt}$$

with L given in (h1), and note that

$$V(0) \leq W(0) \leq W_\varepsilon(0).$$

Let $t_\delta \in \mathbb{T}^+$ be such that

$$t_\delta = \sup \{\delta > 0 \colon V(0) \leq W(0) \text{ implies that } V(t) \leq W_\varepsilon(t) \text{ on } [0, \delta]_\mathbb{T}\}.$$

Evidently, $t_\delta \geq 0$. From Proposition 2.3(I), it follows that V and W are continuous. Next, in view of the continuity of functions V and W, it follows that $t_\delta > 0$ if 0 is right-dense. Moreover, if t_δ is right-dense, then

$$\Delta_g W_\varepsilon(t_\delta) = \Delta_g W(t_\delta) + L\varepsilon e^{Lt_\delta},$$

and if t_δ is right-scattered, then

$$\Delta_g W_\varepsilon(t_\delta) = \Delta_g W(t_\delta) + \varepsilon \frac{e^{L\sigma(t_\delta)} - e^{Lt_\delta}}{\mu(t_\delta)}.$$

We claim that $t_\delta = +\infty$. If not, on the contrary, we have

$$V(t_\delta) \leq W_\varepsilon(t_\delta) \quad \text{and} \quad \sigma(t_\delta) \leq a \quad \text{with some } a \in \mathbb{T}^+.$$

Our assumptions guarantee that

$$\begin{aligned}
\Delta_g V(t_\delta) &\leq -a(t_\delta, V(t_\delta))V(\sigma(t_\delta)) + F(t_\delta, V_{t_\delta}) \\
&\leq -a(t_\delta, W_\varepsilon(t_\delta))W_\varepsilon(\sigma(t_\delta)) + F(t_\delta, (W_\varepsilon)_{t_\delta}) \\
&\leq -a(t_\delta, W(t_\delta))W(\sigma(t_\delta)) + F(t_\delta, (W)_{t_\delta}) \\
&\quad + L(W_\varepsilon(t_\delta) - W(t_\delta)) \\
&= -a(t_\delta, W(t_\delta))W(\sigma(t_\delta)) + F(t_\delta, (W)_{t_\delta}) + L\varepsilon e^{Lt_\delta} \\
&\leq \Delta_g W(t_\delta) + L\varepsilon e^{Lt_\delta}.
\end{aligned}$$

Also, it is easy to see that

$$Le^{Lt_\delta} \leq \frac{e^{L\sigma(t_\delta)} - e^{Lt_\delta}}{\mu(t_\delta)}.$$

Hence, from the above inequality, we obtain

$$\Delta_g V(t_\delta) \leq \Delta_g W_\varepsilon(t_\delta). \tag{3.20}$$

If t_δ is right-dense, then Proposition 2.3(III), together with (3.20), implies that there exists $\eta > 0$ such that

$$W_\varepsilon(t_\delta) - V(t_\delta) \leq W_\varepsilon(t) - V(t) \quad \text{for } t_\delta < t < t_\delta + \eta.$$

This contradicts to the fact that $t_\delta > 0$ is the supremum. Thus, in view of the continuity of V and W, we obtain that the inequality

$$V(t) < W_\varepsilon(t) \quad \text{holds for } t \in \mathbb{T}^+.$$

On the other hand, if t_δ is right-scattered, from Proposition 2.3(II) and (3.20), it follows that

$$\frac{V(\sigma(t_\delta)) - V(t_\delta)}{\mu(t_\delta)} \leq \frac{W_\varepsilon(\sigma(t_\delta)) - W_\varepsilon(t_\delta)}{\mu(t_\delta)},$$

that is,

$$W_\varepsilon(t_\delta) - V(t_\delta) \leq W_\varepsilon(\sigma(t_\delta)) - V(\sigma(t_\delta)).$$

Clearly, $t_\delta < \sigma(t_\delta) \leq a$. This shows that

$$W_\varepsilon(\sigma(t_\delta)) \geq V(\sigma(t_\delta))$$

which contradicts to the fact that t_δ is the supremum. Consequently, we obtain that

$$V(t) < W_\varepsilon(t) \quad \text{holds for } t \in \mathbb{T}^+.$$

Taking limit $\varepsilon \to 0$ yields the desired result on $\mathbb{T}^+$. Next, let $s_\delta \in \mathbb{T}^+$ be such that

$$s_\delta = \inf \{\delta > 0 \colon V(0) \leq W(0) \text{ implies that } V(t) \leq W_\varepsilon(t) \text{ on } [-\delta, 0]_\mathbb{T}\}.$$

Evidently, $s_\delta \leq 0$. In view of the continuity of V and W, it follows that $s_\delta < 0$ if 0 is left-dense. Moreover, $\rho(\sigma(s_\delta)) = s_\delta$, and

$$\Delta_g V(\rho(s_\delta)) \leq -a(\rho(s_\delta), V(\rho(s_\delta))V(s_\delta) + F(\rho(s_\delta), V_{\rho(s_\delta)})$$

and

$$\Delta_g W(\rho(s_\delta)) \geq -a(\rho(s_\delta), W(\rho(s_\delta))W(s_\delta) + F(\rho(s_\delta), W_{\rho(s_\delta)}).$$

We claim that $s_\delta = -\infty$. If not, on the contrary, we have

$$V(s_\delta) \leq W_\varepsilon(s_\delta) \quad \text{and} \quad \rho(s_\delta) \geq b \quad \text{with some } b \in \mathbb{T}^-.$$

The above inequalities, combining with our assumptions, guarantee that

$$\Delta_g V(\rho(s_\delta)) \leq -a(\rho(s_\delta), V(\rho(s_\delta))V(s_\delta) + F(\rho(s_\delta), V_{\rho(s_\delta)})$$
$$\leq -a(\rho(s_\delta), W_\varepsilon(\rho(s_\delta))W_\varepsilon(s_\delta) + F(\rho(s_\delta), (W_\varepsilon)_{\rho(s_\delta)})$$
$$\leq -a(\rho(s_\delta), W(\rho(s_\delta))W(s_\delta) + F(\rho(s_\delta), (W)_{\rho(s_\delta)})$$

$$+ L(W_\varepsilon(\rho(s_\delta))) - W(\rho(s_\delta))$$
$$= -a(\rho(s_\delta), W(\rho(s_\delta))W(s_\delta)$$
$$+ F(\rho(s_\delta), (W)_{\rho(s_\delta)}) + L\varepsilon e^{L\rho(s_\delta)}$$
$$\leq \Delta_g W(\rho(s_\delta)) + L\varepsilon e^{L\rho(s_\delta)}$$
$$\leq \Delta_g W_\varepsilon(\rho(s_\delta)).$$

Hence, s_δ must be left-dense and (3.20) holds with s_δ instead of t_δ. Similarly, we can choose $\eta > 0$ such that

$$W_\varepsilon(s_\delta) - V(s_\delta) \leq W_\varepsilon(t) - V(t) \quad \text{for } s_\delta - \eta < t < s_\delta.$$

This contradicts to the fact that $s_\delta < 0$ is the infimum. Thus, in view of the continuity of V and W, we obtain that the inequality

$$V(t) < W_\varepsilon(t) \quad \text{holds for } t \in \mathbb{T}^-.$$

Again, taking limit $\varepsilon \to 0$ yields the desired result on $\mathbb{T}^-$. This completes the proof. □

Corollary 3.2. *Let* $V, W \in \mathscr{C}_\mathbb{T}$ *be such that*

$$\Delta_g V(t) \leq \Delta_g W(t) \quad \text{for all } t \in \mathbb{T}.$$

Then, $V(0) \leq W(0)$ *implies that*

$$V(t) \leq W(t) \quad \text{for all } t \in \mathbb{T}.$$

We are now in a position to state and prove the existence of solutions to (3.17).

Theorem 3.13. *Assume that the following conditions, besides* (h1), *hold.*

(h2) *The SDE* (3.17) *has the lower solution* V *and the upper solution* W *with* $V, W \in \mathscr{C}_\mathbb{T}$.

(h3) *The function* $F(t, \Phi)$ *is rd-continuous in* $t \in \mathbb{T}$ *and continuous in* $\Phi \in \mathscr{C}_\xi$. *Further,*

$$F(t + \omega, \Phi) = F(t, \Phi) \quad \text{for } \Phi \in \mathscr{C}_\xi$$

and maps bounded sets into bounded sets in $\mathscr{C}_\xi$.

(h4) *The function $a(t, U)$ is rd-continuous and ω-periodic in t, continuous and nondecreasing in U, that is,*

$$a(t, U) \leq a(t, V) \quad \text{for all } t \in \mathbb{T} \quad \text{and} \quad U, V \in \mathscr{C}_\mathbb{T} \quad \text{with } U \leq V.$$

Moreover,

$$a(t, U(t)) \geq \alpha(t) \quad \text{for all } U \in \mathscr{C}_\mathbb{T},$$

where α is regressive, rd-continuous nonnegative function and $k_\alpha > 0$.

Then, (3.17) has the minimal solution Φ and the maximal solution Ψ in $\mathscr{C}_\mathbb{T}$. Moreover, there exist monotone sequences $\{V_n\}$ and $\{W_n\}$ in $\mathscr{C}_\mathbb{T}$ such that $\{V_n\}$ converges to Φ and $\{W_n\}$ converges to Ψ uniformly in $\mathscr{C}_\xi$ as n tends to ∞.

Proof. Let us construct the set integral sequences by

$$\begin{cases} V_{n+1}(t) = \displaystyle\int_t^{t+\omega} v_n(s, t) F(s, V_n(s)) \Delta_g s, \\[4mm] W_{n+1}(t) = \displaystyle\int_t^{t+\omega} w_n(s, t) F(s, W_n(s)) \Delta_g s. \end{cases} \tag{3.21}$$

for $n \in \mathbb{N}_0$. Here, we stipulate

$$V_0(t) = V(t), \quad W_0(t) = W(t), \quad v_n(s, t) = \frac{e_{a_n}(s, t)}{k_{a_n}},$$

$$w_n(s, t) = \frac{e_{b_n}(s, t)}{k_{b_n}}$$

with

$$a_n(t) = a(t, V_n(t)), \quad b_n(t) = a(t, W_n(t)) \quad \text{and} \quad t \in \mathbb{T}.$$

Note that the set-valued functions V_{n+1} and W_{n+1} are all well-defined according to Theorem 2.1. Moreover, in view of Theorems 2.1 and 3.11, we have that

$$V_{n+1}, W_{n+1} \in \mathcal{C}_{\mathrm{rd}}^1(\mathbb{T}, K_c^1)$$

and

$$\begin{cases} \Delta_g V_{n+1}(t) = -a(t, V_n(t)) V_{n+1}(\sigma(t)) + F(t, V_n(t)), \\[2mm] \Delta_g W_{n+1}(t) = -a(t, W_n(t)) W_{n+1}(\sigma(t)) + F(t, W_n(t)). \end{cases} \tag{3.22}$$

Now, we prove that

$$V_0(t) \leq V_1(t) \leq W_1(t) \leq W_0(t) \quad \text{for } t \in \mathbb{T}.$$

Since V_0 is lower solution of (3.17), $V = V_0$ satisfies the first inequality of Definition 3.8. Since a_0 is regressive, rd-continuous, and non-negative in t, it is easy to see $e_{a_0}(t, 0) > 0$ for all $t \in \mathbb{T}$ and

$$\Delta_g[V(t)e_{a_0}(t, 0)] = \Delta_g V(t)e_{a_0}(t, 0) + V(\sigma(t))e_{a_0}^{\Delta}(t, 0)$$
$$= e_{a_0}(t, 0)(\Delta_g V(t) + a_0(t)V(\sigma(t)))$$
$$\leq e_{a_0}(t, 0)F(t, V(t)).$$

Integrating both sides of the above inequality from 0 to ω, and using Theorem 2.1, we obtain

$$\int_0^\omega e_{a_0}(s, 0)F(s, V(s))\Delta_g s \geq \int_0^\omega \Delta_g[V(s)e_{a_0}(s, 0)]\Delta_g s$$
$$= V(\omega)e_{a_0}(\omega, 0) -_g V(0)e_{a_0}(0, 0)$$
$$= V(0)[e_{a_0}(\omega, 0) - e_{a_0}(0, 0)]$$
$$= V(0)k_{a_0}.$$

This implies that

$$V_1(0) = \int_0^\omega v_0(s, 0)F(s, V(s))\Delta_g s \geq V(0).$$

We claim that

$$V(t) \leq V_1(t) \quad \text{for } t \in \mathbb{T}.$$

Setting $U(t) = V_1(t) -_g V_0(t)$, we observe that

$$\Delta_g U(t) = \Delta_g V_1(t) -_g \Delta_g V(t)$$
$$\geq -a(t, V_0(t))V_1(\sigma(t)) + F(t, V_0(t))$$
$$-_g(-a(t, V_0(t))V(\sigma(t)) + F(t, V_0(t)))$$
$$= -a(t, V(t))(V_1(\sigma(t)) -_g V(\sigma(t)))$$
$$= -a_0(t)U(\sigma(t)).$$

That is,

$$\Delta_g U(t) = -a_0(t)U(\sigma(t)) \quad \text{for } t \in \mathbb{T}.$$

This yields that

$$\Delta_g[U(t)e_{a_0}(t,0)] = \Delta_g U(t)e_{a_0}(t,0) + U(\sigma(t))e_{a_0}^{\Delta}(t,0)$$
$$= e_{a_0}(t,0)(\Delta_g U(t) + a_0(t)U(\sigma(t)))$$
$$\geq \theta.$$

From this, it follows that

$$\Delta_g[V_1(t)e_{a_0}(t,0)] \geq \Delta_g[V(t)e_{a_0}(t,0)] \quad \text{for } t \in \mathbb{T}.$$

Now, in virtue of Theorem 1.6 and Corollary 3.2, we have

$$V(t)e_{a_0}(t,0) \leq V_1(t)e_{a_0}(t,0) \quad \text{for } t \in \mathbb{T}.$$

This yields that

$$V(t) \leq V_1(t) \quad \text{for all } t \in \mathbb{T}.$$

Following the method of this proof, it is easy to see that

$$W_1(t) \leq W_0(t) \quad \text{for } t \in \mathbb{T}.$$

Substitute $n = 0$ in (3.22). Then, in view of the monotonicity of F and the inequality $v_0(s,0) \leq w_0(s,0)$, we have

$$V_1(0) = \int_0^\omega v_0(s,0)F(s,V_0(s))\Delta_g s$$
$$\leq \int_0^\omega w_0(s,0)F(s,W_0(s))\Delta_g s$$
$$= W_1(0).$$

Now, we claim that

$$V_1(t) \leq W_1(t) \quad \text{for } t \in \mathbb{T}.$$

Setting $U_1(t) = W_1(t) -_g V^1(t)$, we observe that

$$\Delta_g U_1(t) = \Delta_g W_1(t) -_g \Delta_g V_1(t)$$
$$= -a(t,W_0(t))W_1(\sigma(t)) + F(t,W_0(t))$$
$$\quad -_g[-a(t,V_0(t))V_1(\sigma(t)) + F(t,V_0(t))]$$

$$\geq -a(t, W_0(t))[W_1(\sigma(t)) -_g V_1(\sigma(t))]$$
$$+ F(t, W_0(t)) -_g F(t, V_0(t))$$
$$\geq -b_0(t)U_1(\sigma(t)).$$

That is,

$$\Delta_g U_1(t) \geq -b_0(t)U_1(\sigma(t)) \quad \text{for } t \in \mathbb{T}.$$

This yields that

$$\Delta_g[U_1(t)e_{b_0}(t,0)] = \Delta_g U_1(t)e_{b_0}(t,0) + U_1(\sigma(t))e_{b_0}^\Delta(t,0)$$
$$= e_{b_0}(t,0)(\Delta_g U_1(t) + b_0(t)U_1(\sigma(t)))$$
$$\geq \theta.$$

From this, it follows that

$$\Delta_g[W_1(t)e_{b_0}(t,0)] \geq \Delta_g[V_1(t)e_{b_0}(t,0)] \quad \text{for } t \in \mathbb{T}.$$

Again, Theorem 1.6 and Corollary 3.2, guarantee that

$$V_1(t)e_{b_0}(t,0) \leq W_1(t)e_{b_0}(t,0) \quad \text{for } t \in \mathbb{T}.$$

This yields that

$$V_1(t) \leq W_1(t) \quad \text{for all } t \in \mathbb{T}.$$

Consequently, we arrive at the following conclusion

$$V_0(t) \leq V_1(t) \leq W_1(t) \leq W_0(t) \quad \text{for } t \in \mathbb{T}.$$

Next, we inductively assume

$$V_{j-1}(t) \leq V_j(t) \leq W_j(t) \leq W_{j-1}(t)$$

for $t \in \mathbb{T}$ and $j \geq 1$. We prove that

$$V_j(t) \leq V_{j+1}(t) \leq W_{j+1}(t) \leq W_j(t) \quad \text{for } t \in \mathbb{T}.$$

To end this, substitute $n = j$ in (3.21) and (3.22). Then, by means of the monotonicity of F and a_j, b_j with respect to j, we obtain

$$V_j(0) = \int_0^\omega v_{j-1}(s,0)F(s, V_{j-1}(s))\Delta_g s$$
$$\leq \int_0^\omega v_j(s,0)F(s, V_j(s))\Delta_g s$$
$$= V_{j+1}(0).$$

Repeating the above process, we have

$$\Delta_g[V_{j+1}(t)e_{a_{j+1}}(t,0)] \geq \Delta_g[V_j(t)e_{a_{j+1}}(t,0)] \quad \text{for } t \in \mathbb{T}.$$

It is guaranteed, again, by Theorem 1.6 and Corollary 3.2, that

$$V_j(t)e_{a_{j+1}}(t,0) \leq V_{j+1}(t)e_{a_{j+1}}(t,0).$$

This yields that

$$V_j(t) \leq V_{j+1}(t) \quad \text{for all } t \in \mathbb{T}.$$

Similarly, we show that

$$V_{j+1}(t) \leq W_{j+1}(t) \leq W_j(t) \quad \text{for all } t \in \mathbb{T}.$$

Hence, by induction, we get

$$V_0(t) \leq V_1(t) \leq \cdots \leq V_n(t) \leq \cdots \leq W_n(t) \leq \cdots \leq W_1(t) \leq W_0(t) \tag{3.23}$$

for all $t \in \mathbb{T}$. The continuity of V_n and W_n, for $n = 0, 1, 2, \ldots$, can be seen from their definition. In what follows, we will prove that there exists $\Phi \in \mathscr{C}_\xi$ such that

$$\lim_{n \to \infty} D_\xi[V_n, \Phi] = 0.$$

To this end, we denote

$$\varphi(s) = D^{I_s}[W_0, V_0] + D^{I_s}[V_0, \theta] \quad \text{for } s \in \mathbb{T}^-,$$

where $I_s = [-s, 0]_{\mathbb{T}^-}$. Since $V_0|_{\mathbb{T}^-}, W_0|_{\mathbb{T}^-} \in \mathscr{C}_\xi$, we have

$$\int_{-\infty}^0 \xi(s)\varphi(s)\Delta s = \int_{-\infty}^0 \xi(s)\left[D^{I_s}[W_0, V_0] + D^{I_s}[V_0, \theta]\right]\Delta s < \infty.$$

Now, Lemma 3.2 implies that

$$\begin{aligned}
D^{I_s}[V_n, \theta] &= D^{I_s}[V_n - V_0, V_0] \\
&\leq D^{I_s}[V_n - V_0, \theta] + D^{I_s}[V_0, \theta] \\
&\leq D^{I_s}[W_0 - V_0, \theta] + D^{I_s}[V_0, \theta] \\
&= \varphi(s).
\end{aligned}$$

This shows that the sequence $\{V_n\}$ is uniformly bounded on I_s. In view of Theorem 2.1(iii), we find

$$\Delta_g V_{n+1}(t) = \int_t^{t+\omega} (\ominus a_n)(t) v_n(s,t) F(s, V_n(s)) \Delta_g s$$

$$+ v_n(t+\omega, \sigma(t)) F(t, V_n(t)) -_g v_n(t, \sigma(t)) F(t, V_n(s))$$

$$= (\ominus a_n)(t) V_n(t) + \frac{F(t, V_n(t))}{k_{a_n}}$$

$$\times \left\{ \frac{1}{1 - \mu(t) a_n(t)} e_{a_n}(t+\omega, t) - \frac{1}{1 + \mu(t) a_n(t)} \right\}$$

$$= (\ominus a_n)(t) F(t, V_n(t)) + \frac{F(t, V_n(t))}{1 + \mu(t) a_n(t)}.$$

In virtue of (h3), F is bounded on $\{(t, V_n): t \in I_s,\ n \in \mathbb{N}_0\}$ and (h4) shows $\{a_n\}$ is also bounded, i.e., there exist $M_1, M_2 > 0$ such that

$$D[F(t, V_n(t)), \theta] \le M_1 \quad \text{and} \quad |a_n(t)| < M_2$$

for $t \in I_s$ and $n \in \mathbb{N}_0$. Thus,

$$D[\Delta_g V_{n+1}(t), \theta] \le M_2 M_1 + M_1.$$

This implies that $\{V^n\}$ is equicontinuous on I_s. From Theorem A.1, it follows that the sequence $\{V_n\}$ converges uniformly to Φ on I_s, i.e.,

$$\lim_{n \to \infty} D^{I_s}[V_n, \Phi] = 0.$$

Now, Theorem 3.9(ii) guarantees that

$$\lim_{n \to \infty} D_\xi[V_n, \Phi] = 0$$

and Theorem 3.9(iii) guarantees $\Phi \in \mathscr{C}_\xi$. Similarly, there exists $\Psi \in \mathscr{C}_\xi$ such that

$$\lim_{n \to \infty} D_\xi[W_n, \Psi] = 0.$$

On the analogy of the above process, one can check that

$$D^\Omega[V_n, \Phi] \to 0 \quad \text{and} \quad D^\Omega[W_n, \Psi] \to 0 \quad \text{as } n \to \infty,$$

where $\Omega = [0, \omega]_{\mathbb{T}}$. Theorem 3.10(i) guarantees that

$$\Phi(t), \Psi(t) \in \mathscr{C}_\xi \quad \text{for any } t \in \Omega,$$

and

$$\bar{\Phi}, \bar{\Psi} \in \mathscr{C}(\Omega, K_c^1) \quad \text{with } \bar{\Phi}(t) = \Phi(t), \quad \bar{\Psi}(t) = \Psi(t).$$

Letting $n \to \infty$, by means of (3.21) and the continuity, one has

$$\Phi(t) = \int_t^{t+\omega} v(t, s) F(s, \Phi(s)) \Delta_g s, \quad \text{and}$$

$$\Psi(t) = \int_t^{t+\omega} w(t, s) F(s, \Psi(s)) \Delta_g s \quad \text{for } t \in \mathbb{T}^- \cup \Omega,$$

(3.24)

where

$$v(s, t) = \frac{e_a(s, t)}{k_a}, \quad w(s, t) = \frac{e_b(s, t)}{k_b}$$

with

$$a(t) = a(t, \Phi(t)) \quad \text{and} \quad b(t) = a(t, \Psi(t)).$$

For $t > \omega$, we write $t = t_1 + n\omega$, where $t_1 \in \Omega$ and $n \in \mathbb{N}$, and define $\Phi(t) = \Phi(t_1)$, $\Psi(t) = \Psi(t_1)$. Then, both Φ and Ψ are ω-periodic such that

$$\lim_{n \to \infty} D_\xi[V_n, \Phi] = 0 \quad \text{and} \quad \lim_{n \to \infty} D_\xi[W_n, \Psi] = 0$$

hold uniformly and (3.24) is satisfied on $\mathbb{T}$. Consequently, it is easy to see $\Phi, \Psi \in \mathscr{C}_{\mathbb{T}}$.

Next, in view of Lemma 3.1, both

$$D_\xi[V_n(t), \Phi(t)] \quad \text{and} \quad D_\xi[W_n(t), \Psi(t)]$$

converge uniformly to 0 as $n \to \infty$. Using this, and passing to the limit $n \to \infty$ in (3.22), we find that both Φ and Ψ are solutions of (3.17). Also, by means of (3.21), we easily get

$$V_0(t) \le \Phi(t) \le \Psi(t) \le W_0(t) \quad \text{for } t \in \mathbb{T}.$$

Now, it remains to show that Φ and Ψ are the minimal and maximal solutions of (3.17), respectively. To end this, we assume that U is any solution of (3.17) satisfying

$$V_0(t) \le U(t) \le W_0(t) \quad \text{for every } t \in \mathbb{T}.$$

Thus, we have to prove that

$$V_0(t) \le \Phi(t) \le U(t) \le \Psi(t) \le W_0(t) \quad \text{for } t \in \mathbb{T}. \tag{3.25}$$

We first prove that

$$V_1(t) \le U(t) \le W_1(t) \quad \text{for each } t \in \mathbb{T}.$$

By Theorem 3.11, we see that U is an ω-periodic solution of (3.17) if and only if

$$U(t) = \int_t^{t+\omega} v_a(s,t) F(s, U(s)) \Delta_g s,$$

where

$$v_a(s,t) = \frac{e_a(s,t)}{k_a} \quad \text{with } a(t) = a(t, U(t)).$$

Using $V_0(t) \le U(t)$, together with (3.21) and the monotonicity of F and a, we find that

$$U(0) = \int_0^\omega v_a(s, 0) F(s, U(s)) \Delta s$$

$$\ge \int_0^\omega v_0(s, 0) F(s, V_0(s)) \Delta s$$

$$= V_1(0).$$

Moreover, setting

$$Z(t) = U(t) -_g V_1(t) \quad \text{for } t \in \mathbb{T},$$

we observe that

$$\begin{aligned}
\Delta_g Z(t) &= \Delta_g U(t) -_g \Delta_g V_1(t) \\
&= -a(t, U(t)) U(\sigma(t)) \\
&\quad + F(t, U(t)) -_g [-a(t, V_0(t)) V_1(\sigma(t)) + F(t, V_0(t))] \\
&\ge -a(t, U(t)) [U(\sigma(t)) -_g V_1(\sigma(t))] + F(t, U_t) -_g F(t, V_0(t)) \\
&\ge -a(t) Z(\sigma(t)).
\end{aligned}$$

This yields that

$$\Delta_g[Z(t)e_a(t,0)] = \Delta_g Z(t)e_a(t,0) + Z(\sigma(t))e_a^\Delta(t,0)$$
$$= e_a(t,0)(\Delta_g Z(t) + a(t)Z(\sigma(t)))$$
$$\geq \theta.$$

From this, it follows that

$$\Delta_g[U(t)e_a(t,0)] \geq \Delta_g[V_1(t)e_a(t,0)] \quad \text{for } t \in \mathbb{T}.$$

In virtue of Theorem 1.6 and Corollary 3.2, we have

$$V_1(t)e_a(t,0) \leq U(t)e_a(t,0) \quad \text{for } t \in \mathbb{T}$$

This yields that

$$V_1(t) \leq U(t) \quad \text{for all } t \in \mathbb{T}.$$

With the analogous method, we get

$$U(t) \leq W_1(t) \quad \text{for each } t \in \mathbb{T}.$$

Moreover, the induction principle immediately implies that

$$V_n(t) \leq U(t) \leq W_n(t)$$

for $n \in \mathbb{N}_0$ and each $t \in \mathbb{T}$. Taking the limit $n \to \infty$, inequalities (3.25) is established, and this completes the proof. □

Remark 3.3. The systematical explorations of the periodic solutions of set dynamic equations with infinite delay on time scales reveal that, although, to the best of our knowledge, no other work has been found to deal with the existence of periodic solutions for set functional differential equations or set functional difference equations with infinite delay, it is unnecessary to prove results for differential equations and separately again for their discrete analogues when the sets under consideration are single-valued mappings. We can, of course, unify such problems of set functional (differential and difference) equations in the framework of set dynamic equations on time scales, too.

3.4   Impulsive Problems

In this section, we study a class of nonlinear impulsive set dynamic equations on time scales. We establish certain criteria for the existence of solutions to such model. Our approach is based on exponential dichotomy of homogeneous SDEs and fixed point theory. The existence conditions given in terms of exponential dichotomy are more concise and explicit than conditions obtained by using the Lyapunov functions.

3.4.1   *Solvability of impulsive SDEs*

We emphasize that $\mathbb{T}^+ = \{t \in \mathbb{T} : t \geq 0\}$ and for $k \in \mathbb{N}$,

$$\mathcal{C}_k = \Big\{ F \colon (t_{k-1}, t_k] \times \mathcal{C}_{\mathrm{rd}} \to K_c^n \colon \text{at each point } (t, U) \in (t_{k-1}, t_k]$$
$$\times \, \mathcal{C}_{\mathrm{rd}}, F \text{ is continuous at right-dense point and has the}$$
$$\text{limit at left-dense point} \Big\}.$$

Moreover, let $U_{t_k^+} = U(t_k^+)$ represent the right limit of U at t_k if t_k is right-dense and $U_{t_k^+} = U(\sigma(t_k))$ if t_k is right-scattered for $k \in \mathbb{N}$. Let

$$PC[\mathbb{T}^+ \times \mathcal{C}_{\mathrm{rd}}, K_c^n]$$

$$= \Big\{ F \colon \mathbb{T}^+ \times \mathcal{C}_{\mathrm{rd}} \to K_c^n \colon F \in \mathcal{C}_k \text{ and}$$

$$\lim_{t \to t_k^+} F(t, U) = F(t_k^+, U) \text{ exists for } U \in \mathcal{C}_{\mathrm{rd}} \text{ and } k \in \mathbb{N} \Big\},$$

$$BC = BC[\mathbb{T}^+, K_c^n] = \Big\{ U \in PC[\mathbb{T}^+, K_c^n] \colon \|U(t)\| \text{ is bounded in } \mathbb{T}^+ \Big\},$$

and

$$PC^1 = PC^1[\mathbb{T}^+, K_c^n] = \{ U \in BC \colon U \text{ is } \Delta_g\text{-differentiable in}$$
$$(t_{k-1}, t_k), \ k \in \mathbb{N} \}.$$

Note that (BC, D_0) is a complete metric space, when it is endowed with the distance

$$D_0[U, V] = \sup_{t \in \mathbb{T}^+} D[U(t), V(t)].$$

Consider the impulsive set dynamic equation (ISDE)

$$\begin{cases} \Delta_g U(t) = A(t)U(t) + F(t, U(t)), & t \in \mathbb{T}^+, \ t \neq t_k, \ k \in \mathbb{N}_0, \\ U(t_k^+) = J_k U(t_k), & t_k \in \mathbb{T}^+, \ k \in \mathbb{N}_0, \\ U(t_0) = U_0 \in K_c^\tau, & t_0 \in \mathbb{T}^+, \end{cases} \quad (3.26)$$

where $F \in PC[\mathbb{T}^+ \times \mathcal{C}_{\mathrm{rd}}, K_c^n]$, $J_k \colon K_c^n \to K_c^n$ is a continuous linear operator, that is, for any $U, V \in K_c^n$, and $a, b \in \mathbb{R}$, one has

$$J_k(aU \pm_g bV) = aJ_k(U) \pm_g bJ_k(V),$$

and $\{t_k\} \subset \mathbb{T}^+$ is a sequence of points such that $0 \leq t_0 < t_1 < \cdots < t_k < \cdots$ and

$$\lim_{k \to \infty} t_k = \infty.$$

By a solution of ISDE (3.26), we mean that a set-valued function $U \in PC^1[\mathbb{T}^+, K_c^n]$ that satisfies (3.26).

3.4.2 *Exponential dichotomy*

To explore the existence of solutions to the impulsive set dynamic equations, we introduce the exponential dichotomy of SDEs. We define the product of a matrix and a subset in $\mathbb{R}^n$ as follows:

$$A(t)U = \{A(t)u \colon u \in U\} \quad \text{for } t \in \mathbb{T}, \quad (3.27)$$

for an $m \times n$-matrix-valued function A on $\mathbb{T}$ and subset $U \subset \mathbb{R}^n$.

By analogous to ordinary dynamic calculus, we have the following useful results.

Theorem 3.14. *Let $A \in \mathcal{R}_n$, $t_0 \in \mathbb{T}$, $U_0 \in K_c^n$, and $F \colon \mathbb{T} \to K_c^n$ be an rd-continuous. Then, the set dynamic initial value problem*

$$\begin{cases} \Delta_g U(t) = A(t)U(t) + F(t), & t \in \mathbb{T}, \\ U_{t_0} = U_0 \end{cases} \quad (3.28)$$

has a unique solution $\mathscr{U} \colon \mathbb{T} \to K_c^n$ given by

$$\mathscr{U}(t, t_0, U_0) = e_A(t, t_0)U_0 + \int_{t_0}^t e_A(t, \sigma(s))F(s)\Delta_g s. \quad (3.29)$$

Theorem 3.15. *Let $A \in \mathcal{R}_n$, $t_0 \in \mathbb{T}$, $U_0 \in K_c^n$, and $F \colon \mathbb{T} \to K_c^n$ be rd-continuous. Then, the set dynamic initial value problem*

$$\begin{cases} \Delta_g U(t) = A(t) U^\sigma(t) + F(t), & t \in \mathbb{T}, \\ U_{t_0} = U_0 \end{cases} \tag{3.30}$$

has a unique solution $\mathscr{U}^* \colon \mathbb{T} \to K_c^n$ given by

$$\mathscr{U}^*(t, t_0, U_0) = e_{\ominus A^*}(t, t_0) U_0 + \int_{t_0}^{t} e_{\ominus A^*}(t, \sigma(s)) F(s) \Delta_g s. \tag{3.31}$$

In what follows, by means of $e_A(t, s) U_s$ we denote the unique solution $\mathscr{U}(t, s, U_s)$ of the linear homogeneous set dynamic equation

$$\Delta_g U(t) = A(t) U(t), \quad t \in \mathbb{T}^+ \tag{3.32}$$

with initial condition

$$U(s) = U_s \in K_c^n, \tag{3.33}$$

where initial point $s \in \mathbb{T}^+$ is fixed. We call $e_A(t, s)$ the fundamental matrix of (3.32) with initial condition (3.33).

Definition 3.10. The SDE (3.32) is said to admit an exponential dichotomy on $\mathbb{T}$ provided there exist positive constants k and a, and a continuous projection (matrix) $P(s)$ (i.e., $P^2(s) = P(s)$) on $\mathbb{R}^n$ such that

$$|e_A(t, s) P(s) e_{\ominus A^*}^*(\sigma(t'), s)| \le k e_{\ominus a}(t, \sigma(t')) \quad \text{for } t \ge \sigma(t') \ge s,$$

$$|e_A(t, s)(\mathcal{I} - P(s)) e_{\ominus A^*}^*(\sigma(t'), s)| \le k e_{\ominus a}(t, \sigma(t')) \quad \text{for } s \ge t \ge t',$$

$$\tag{3.34}$$

where $t', t \in \mathbb{T}^+$, and $e_A(t, s)$ is the fundamental matrix of (3.32) and $|B|$ is the norm of the $n \times n$-matrix $B = (b_{ij})_{n \times n}$, given by

$$|B| = \left(\sum_{i=1}^{n} \sum_{j=1}^{n} |b_{ij}|^2 \right)^{1/2}.$$

Theorem 3.16. *Let A be an $n \times n$-matrix-valued function such that*

$$e_A(t, 0) = \sum_{i=0}^{n-1} r_{i+1}(t) P_i,$$

where P_i $(i = 0, 1, \ldots, n - 1)$ *are constant* $n \times n$*-matrices and* $r(t) = (r_1(t), r_2(t), \ldots, r_n(t))^T$ *is the solution of dynamic initial value problem*

$$r^\Delta = \begin{pmatrix} -\lambda_1(t) & 0 & 0 & \cdots & 0 & 0 \\ 1 & -\lambda_2(t) & 0 & \cdots & 0 & 0 \\ 0 & 1 & -\lambda_3(t) & \cdots & 0 & 0 \\ \vdots & \vdots & \vdots & \ddots & \vdots & \vdots \\ 0 & 0 & 0 & \cdots & 1 & -\lambda_n(t) \end{pmatrix} r, \quad r(0) = \begin{pmatrix} 1 \\ 0 \\ 0 \\ \vdots \\ 0 \end{pmatrix}.$$

If $-\lambda_i \in \mathcal{R}_1^+$ *for* $i = 1, 2, \ldots, n$*, and*

$$p = \min_{1 \le i \le n} \left\{ \inf_{t \in \mathbb{T}} \lambda_i(t) \right\} > 0,$$

then (3.32) *admits an exponential dichotomy on* $\mathbb{T}$.

Proof. In fact, it is easy to check that

$$\begin{cases} r_1(t) = e_{-\lambda_1}(t, 0), \\ r_{i+1}(t) = \int_0^t e_{-\lambda_{i+1}}(t, \sigma(\tau)) r_i(\tau) \Delta\tau, \quad i = 1, 2, \ldots, n - 1. \end{cases}$$

For any $s, t \in \mathbb{T}^+ =: \mathbb{T} \cap \mathbb{R}^+$ with $s < t$, consider the following two cases.

Case I. If $\tau \in [s, t)_{\mathbb{T}}$ is right-scattered, then $\mu(\tau) > 0$ and

$$0 < 1 - \mu(\tau)\lambda_i(\tau)$$
$$\le 1 - p\mu(\tau)$$
$$< 1 - \frac{p\mu(\tau)}{1 + p\mu(\tau)}$$
$$= \frac{1}{1 + p\mu(\tau)}.$$

That is,

$$1 - \mu(\tau)\lambda_i(\tau) < \frac{1}{1 + p\mu(\tau)}.$$

This implies that

$$\int_s^t \frac{\mathrm{Log}(1 - \lambda_i \mu(\tau))}{\mu(\tau)} \Delta\tau \le \int_s^t \frac{\mathrm{Log}\left(\frac{1}{1+p\mu(\tau)}\right)}{\mu(\tau)} \Delta\tau.$$

Therefore,

$$e_{-\lambda_i}(t, s) = \exp\left\{ \int_s^t \frac{\mathrm{Log}(1 - \lambda_i \mu(\tau))}{\mu(\tau)} \Delta\tau \right\}$$

$$\le \exp\left\{ \int_s^t \frac{\mathrm{Log}\left(\frac{1}{1+p\mu(\tau)}\right)}{\mu(\tau)} \Delta\tau \right\}$$

$$= e_{\ominus p}(t, s).$$

Case II. If $\tau \in [s, t)_{\mathbb{T}}$ is right-dense, then $\mu(\tau) = 0$, and we have

$$e_{-\lambda_i}(t, s) = \exp\left\{ \int_s^t -\lambda_i(\tau)\Delta\tau \right\}$$

$$\le \exp\left\{ \int_s^t -p\,\Delta\tau \right\}$$

$$= e_{\ominus p}(t, s).$$

From the above two cases, we get

$$r_1(t) = e_{-\lambda_1}(t, 0) \le e_{\ominus p}(t, 0),$$

if we set $s = 0$ and $i = 1$. Moreover, since $\mu(t) \le \frac{1}{p}$ for all $t \in \mathbb{T}^+$, we have

$$r_2(t) = \int_0^t e_{-\lambda_2}(t, \sigma(\tau)) r_1(\tau)\Delta\tau$$

$$\le \int_0^t e_{\ominus(-\lambda_2)}(\sigma(\tau), t) e_p(0, \tau)\Delta\tau$$

$$= \int_0^t [1 - \lambda_2(\tau)\mu(\tau)]^{-1} e_{\ominus(-\lambda_2)}(\tau, t) e_p(0, \tau)\Delta\tau$$

$$= \int_0^t [1 - \lambda_2(\tau)\mu(\tau)]^{-1} e_{-\lambda_2}(t, \tau) e_p(0, \tau)\Delta\tau$$

$$\leq \int_0^t [1 - \lambda_2(\tau)\mu(\tau)]^{-1} e_{\ominus p}(t,\tau) e_{\ominus p}(\tau,0)\Delta\tau$$

$$= e_{\ominus p}(t,0) \int_0^t [1 - \lambda_2(\tau)\mu(\tau)]^{-1}\Delta\tau.$$

Thus, we infer, recursively, that

$$r_{i+1}(t) \leq e_{\ominus p}(t,0) \quad \text{for } i = 2,3,\ldots,n-2.$$

Hence,

$$|e_A(t,0)| \leq \sum_{i=0}^{n-1} |r_{i+1}(t)||P_i|$$

$$\leq nM e_{\ominus p}(t,0),$$

where $M = \max\{|P_i| : 0 \leq i \leq n-1\}$. By means of Theorem 1.9(ii), we have

$$e^*_{\ominus A^*}(s,0) = e_A(0,s).$$

Thus, for $s \in \mathbb{T}$,

$$|e^*_{\ominus A^*}(s,0)| = |e_A(0,s)|$$

$$\leq nM e_{\ominus p}(0,s).$$

Now, taking the projection $P \equiv \mathcal{I}$, we obtain

$$|e_A(t,0)Pe^*_{\ominus A^*}(s,0)| \leq n^2 M^2 e_{\ominus p}(t,0)e_{\ominus p}(0,s)$$

$$\leq k e_{\ominus p}(t,\sigma(s)), \quad s,t \in \mathbb{T},$$

where $k = n^2 M^2$. Therefore, in view of Definition 3.10, SDE (3.32) admits an exponential dichotomy on $\mathbb{T}$ with $P \equiv \mathcal{I}$. This completes the proof. $\qquad\square$

Remark 3.4. Suppose that the $\lambda_1, \lambda_2, \ldots, \lambda_n$ given in Theorem 3.16 are eigenvalues of constant $n \times n$-matrix $A \in \mathcal{R}_n$ and are constants. Let $A(t) \equiv -A$ and assume $\lambda_i > 0$ and $-\lambda_i \in \mathcal{R}_1^+$ for $i = 1,2,\ldots,n$. In this case, the Putzer algorithm, given in Theorem 1.11, guarantees that the P-matrices $P_0, P_1, \ldots, P_n$ are recursively defined by $P_0 = \mathcal{I}$ and

$$P_{i+1} = (-A + \lambda_{i+1}\mathcal{I})P_i, \quad 0 \leq i \leq n-1.$$

Then, SDE (3.32) still admits an exponential dichotomy with fundamental matrix $e_A(t, 0)$ given as in Theorem 3.16.

For further discussion, we need the following hypotheses:

(H1) For $t, t' \in \mathbb{T}^+$, the equality

$$e_A(t, s)(\mathcal{I} - P(s))e^*_{\ominus A^*}(\sigma(t'), s)F(t')$$
$$= e_A(t, s)e^*_{\ominus A^*}(\sigma(t'), s)F(t') -_g e_A(t, s)P(s)e^*_{\ominus A^*}(\sigma(t'), s)F(t')$$

holds. (In general, $(a - b)U \neq aU -_g bU$ for $a, b \in \mathbb{R}$ and $U \in K_c^n$).

(H2) There exists a constant $0 < \chi < 1$ such that

$$e_a(\sigma(t_k), \sigma(t_{k+1})) \leq \chi$$

for $t_k \in \mathbb{T}^+$, $k = 0, 1, 2, \ldots$, where a is given as in Definition 3.10.

Let

$$\mathcal{H}_n = \left\{\mathcal{H} = \{H_m\} \colon H_m \in K_c^n \text{ and } \sup\{\|H_m\| \colon m \in \mathbb{N}\} < \infty\right\},$$

$$D_\infty[\mathcal{H}^1, \mathcal{H}^2] = \sup\left\{D[H_m^1, H_m^2] \colon m \in \mathbb{N}\right\}$$

with

$$\mathcal{H}^i = \{H_m^i\} \in \mathcal{H}_n \ (i = 1, 2),$$

and

$$\|\mathcal{H}\|_\infty = D_\infty[\mathcal{H}, \Theta], \text{ where } \Theta \text{ is the zero element of } \mathcal{H}_n.$$

Note that $(\mathcal{H}_n, D_\infty)$ is a complete metric space.

Theorem 3.17. *Let $A \in \mathcal{R}_n$ be an $n \times n$-matrix-valued function such that the SDE (3.32) with $s = t_0$ admit the exponential dichotomy on $\mathbb{T}^+$ with positive constants k, a and the projection P.*

Also, the hypotheses (H1) and (H2) hold. If $F \in BC[\mathbb{T}^+, K_c^n]$ and $\mathscr{H} = \{H_m\} \in \mathscr{H}_n$, then the linear ISDE

$$\begin{cases} \Delta_g U(t) = A(t)U(t) + F(t), & t \neq t_k, \ (k \in \mathbb{N}_0), \\ U(t_k^+) = J_k U(t_k) + H_k, & k \in \mathbb{N}_0, \\ U(t_0) = U_0 \end{cases} \quad (3.35)$$

has a unique bounded solution $\mathscr{U}(F, \mathscr{H})$ on $\mathbb{T}^+$ satisfying

$$\mathscr{U}(F, \mathscr{H})(t) = \begin{cases} \mathscr{U}_0(t, t_0, U_0) & \text{for } t \in [0, t_0]_{\mathbb{T}+} \\ \mathscr{U}_1(t) & \text{for } t \in (t_0, \infty)_{\mathbb{T}+}, \end{cases} \quad (3.36)$$

where

$$\begin{aligned} \mathscr{U}_1(t) = &\int_{t_0}^{t} e_A^0(t) P e_{\ominus A^*}^0(\sigma(\tau)) F(\tau) \Delta_g \tau \\ &-_g \int_{t}^{\infty} e_A^0(t) P_1 e_{\ominus A^*}^0(\sigma(\tau)) F(\tau) \Delta_g \tau \\ &+ \sum_{t_j < t} e_A^0(t) P e_{\ominus A^*}^0(t_j^+) H_j -_g \sum_{t_j \geq t} e_A^0(t) P e_{\ominus A^*}^0(t_j^+) H_j \end{aligned} \quad (3.37)$$

with $t \in \mathbb{T}^+$, $e_A^0(t) = e_A(t, t_0)$, $e_{\ominus A^*}^0(t) = e_{\ominus A^*}^*(t, t_0)$, $P = P(t_0)$, and $P_1 = (\mathcal{I} - P)$.

Proof. To show that $\mathscr{U} : BC \times \mathscr{H}_n \to BC$ is continuous and $\mathscr{U}(F, \mathscr{H}) \in PC^1$, we shall first estimate the $\| \cdot \|$ of the addends in (3.37) for $t \in (t_0, \infty)_{\mathbb{T}+}$. Let M be a positive constant such that

$$\|F(t)\| \leq M \quad \text{for all } t \in \mathbb{T}^+.$$

By Definition 3.10 and Proposition 2.5(i6), we have

$$\begin{aligned} \left\| \int_{t_0}^{t} e_A^0(t) P e_{\ominus A^*}^0(\sigma(\tau)) F(\tau) \Delta_g \tau \right\| \\ \leq \int_{t_0}^{t} \|e_A^0(t) P e_{\ominus A^*}^0(\sigma(\tau)) F(\tau)\| \Delta \tau \\ = \int_{t_0}^{t} |e_A^0(t) P e_{\ominus A^*}^0(\sigma(\tau))| \, \|F(\tau)\| \Delta \tau \end{aligned}$$

$$\leq k \int_{t_0}^{t} e_{\ominus a}(t, \sigma(\tau)) \|F(\tau)\| \Delta \tau$$

$$= k e_{\ominus a}(t, 0) \int_{t_0}^{t} e_{\ominus a}(0, \sigma(\tau)) \|F(\tau)\| \Delta \tau$$

$$\leq kM \int_{t_0}^{t} e_{\ominus a}(0, \sigma(\tau)) \Delta \tau$$

$$\leq \frac{kM}{a},$$

and

$$\left\| \int_{t}^{\infty} e_A^0(t) P_1 e_{\ominus A^*}^0(\sigma(\tau)) F(\tau) \Delta_g \tau \right\|$$

$$\leq \int_{t}^{\infty} \| e_A^0(t) P_1 e_{\ominus A^*}^0(\sigma(\tau)) F(\tau) \| \Delta \tau$$

$$= \int_{t}^{\infty} \left| e_A^0(t) P_1 e_{\ominus A^*}^0(\sigma(\tau)) \right| \|F(\tau)\| \Delta \tau$$

$$\leq k \int_{t}^{\infty} e_{\ominus a}(\sigma(\tau), t) \|F(\tau)\| \Delta \tau$$

$$\leq kM e_a(t, 0) \int_{t}^{\infty} e_a(0, \sigma(\tau)) \Delta \tau$$

$$\leq \frac{kM}{a},$$

and analogously (noting that $e_{\ominus A^*}^0(t_j^+) = e_{\ominus A^*}^0(\sigma(t_j))$, we obtain

$$\left\| \sum_{t_j < t} e_A^0(t) P e_{\ominus A^*}^0(t_j^+) H_j \right\| \leq \sum_{t_j < t} k e_{\ominus a}(t, \sigma(t_j)) \|H_j\|$$

$$\leq k \|\mathscr{H}\|_\infty \sum_{t_j < t} e_{\ominus a}(t, \sigma(t_j)), \qquad (3.38)$$

and

$$\left\| \sum_{t \leq t_j} e_A^0(t) P_1 e_{\ominus A^*}^0(t_j^+) H_j \right\| \leq \sum_{t \leq t_j} k e_{\ominus a}(\sigma(t_j), t) \|H_j\|$$

$$\leq k \|\mathscr{H}\|_\infty \sum_{t \leq t_j} e_{\ominus a}(\sigma(t_j), t). \qquad (3.39)$$

Next, assuming $t_m \leq t \leq t_{m+1}$ for some $m \in \mathbb{N}$, from (H2), it follows that

$$\sum_{t_j < t} e_{\ominus a}(t, \sigma(t_j)) \leq \sum_{j=0}^{m} e_{\ominus a}(\sigma(t_m), \sigma(t_j))$$

$$= \sum_{j=0}^{m} e_a(\sigma(t_j), \sigma(t_{j+1})) \cdots e_a(\sigma(t_{m-1}), \sigma(t_m))$$

$$\leq \sum_{j=0}^{m} \chi^{m-j}$$

$$\leq \frac{1}{1 - \chi},$$

and

$$\sum_{t \leq t_j} e_{\ominus a}(\sigma(t_j), t)$$

$$\leq \sum_{j=0}^{\infty} e_{\ominus a}(\sigma(t_j), \sigma(0))$$

$$= \sum_{j=0}^{\infty} e_a(\sigma(0), \sigma(t_1)) e_a(\sigma(t_1), \sigma(t_2)) \cdots e_a(\sigma(t_{j-1}), \sigma(t_j))$$

$$\leq e_a(\sigma(0), \sigma(t_1)) \sum_{j=0}^{\infty} \chi^{j-1}$$

$$= \frac{1}{1 - \chi} e_a(\sigma(0), \sigma(t_1)).$$

Now, substituting the above two inequalities into (3.38) and (3.39), respectively, we have

$$\left\| \sum_{t_j < t} e_A^0(t) P e_{\ominus A^*}^0(t_j^+) H_j \right\| \leq \frac{k}{1 - \chi} \|\mathscr{H}\|_\infty,$$

$$\left\| \sum_{t \leq t_j} e_A^0(t) P_1 e_{\ominus A^*}^0(t_j^+) H_j \right\| \leq \frac{k e_a(\sigma(0), \sigma(t_1))}{1 - \chi} \|\mathscr{H}\|_\infty.$$

(3.40)

From the above inequalities, it follows that the set-valued function $\mathscr{U}(F, \mathscr{H})$ given in (3.36) is bounded on $(t_0, \infty)_{\mathbb{T}^+}$. If $t \in [0, t_0]_{\mathbb{T}^+}$, from the exponential dichotomy, it follows that $e_A^0(t)$ is bounded, that is,

$$\|e_A^0(t)\| \leq E \quad \text{for some } E > 0.$$

Thus,

$$\|\mathscr{U}(F, \mathscr{H})(t)\| = \|\mathscr{U}_0(t, t_0, U_0)\|$$

$$= \left\| e_A^{\mathsf{C}}(t) \left(U_0 + \int_{t_0}^{t} e_A(t_0, \sigma(\tau)) F(\tau) \Delta_g \tau \right) \right\|$$

$$\leq |e_A^0(t)| \|U_0\| + |e_A^0(t)| \int_0^{t_0} \|e_A(t_0, \sigma(\tau)) F(\tau)\| \Delta_g \tau$$

$$\leq E(|U_0\| + M E_0),$$

where

$$E_0 = \int_0^{t_0} |e_A(t_0, \sigma(\tau))| \Delta \tau.$$

This implies that $\mathscr{U}(F, \mathscr{H})$ is bounded on $[0, t_0]_{\mathbb{T}^+}$. Consequently, $\mathscr{U}(F, \mathscr{H})$ is bounded on $\mathbb{T}^+$. The continuity of $\mathscr{U}(F, \mathscr{H})$ for $t \neq t_k$ ($k \in \mathbb{N}_0$) and the existence of the limit values $\mathscr{U}(F, \mathscr{H})(t_k^+)$ ($k \in \mathbb{N}_0$) are immediately verified by Proposition 2.5(i6). Thus, $\mathscr{U}(F, \mathscr{H}) \in BC$. Next, we verify that $\mathscr{U}(F, \mathscr{H}) \in PC^1$. Differentiating (3.37) for $t_k \neq t \in (t_0, \infty)_{\mathbb{T}^+}$ ($k \in \mathbb{N}_0$) and taking into account Propositions 2.4(d1) and (d2), Proposition 2.5(i7) and Theorem 2.1(iii), we obtain

$$\Delta_g \mathscr{U}(F, \mathscr{H})(t) = e_A^0(\sigma(t)) P e_{\ominus A^*}^0(\sigma(t)) F(t)$$

$$- e_A^0(\sigma(t)) (\mathcal{I} - P) e_{\ominus A^*}^0(\sigma(t)) F(t)$$

$$- \int_{t_0}^{t} A(t) e_A^0(t) P e_{\ominus A^*}^0(\sigma(t)) F(t) \Delta_g \tau$$

$$-_g \int_t^{\infty} A(t) e_A^0(t) (\mathcal{I} - P) e_{\ominus A^*}^0(\sigma(t)) F(t) \Delta_g \tau$$

$$+ \sum_{t_j < t} A(t) e_A^0(\sigma(t)) P e_{\ominus A^*}^0(t_j^+) H_j$$

$$-g \sum_{t_j \geq t} A(t) e_A^0(\sigma(t)) P_1 e_{\ominus A^*}^0(t_j^+) H_j$$

$$= F(t) + A(t)\mathscr{U}(F,\mathscr{H})(t).$$

That is,

$$\Delta_g \mathscr{U}(F,\mathscr{H})(t) = F(t) + A(t)\mathscr{U}(F,\mathscr{H})(t) \text{ for } t \neq t_k \text{ and } t \in (t_0,\infty)_{\mathbb{T}+}.$$
$$(3.41)$$

In the case of $t \in [0,t_0)_{\mathbb{T}+}$, observing

$$\mathscr{U}(t,t_0,U_0) = e_A^0(t)\left(U_0 + \int_{t_0}^t e_A(t_0,\sigma(\tau))F(\tau)\Delta_g\tau\right),$$

we have

$$\Delta_g \mathscr{U}(F,\mathscr{H})(t) = \Delta_g \mathscr{U}_0(t,t_0,U_0)$$

$$= A(t)e_A^0(t)\left(U_0 + \int_{t_0}^t e_A(t_0,\sigma(\tau))F(\tau)\Delta_g\tau\right)$$

$$+ e_A^0(\sigma(t))e_A(t_0,\sigma(t))F(t)$$

$$= A(t)\mathscr{U}(t,t_0,U_0) + F(t)$$

$$= A(t)\mathscr{U}(F,\mathscr{H}) + F(t).$$

This implies that $\mathscr{U}(F,\mathscr{H}) \in PC^1$. Now, it remains to prove that $\mathscr{U}(F,\mathscr{H})$ satisfies (3.35). From (3.41), we have proved that the first equation of (3.35) holds for $t \neq t_k$. Consider system (3.26) with $F \equiv \theta$ and $U_0 = V \in K_c^n$. Evidently, this system has a unique solution $e_A^0(t)V$. According to the impulsive condition, we obtain

$$e_A^0(t_k^+)V = J_k e_A^0(t_k)V \quad \text{for } V \in K_c^n, \ k \in \mathbb{N}_0.$$

Taking into account this and combining with (H1), we obtain

$$\mathscr{U}(F,\mathscr{H})(t_k^+)$$

$$= \int_{t_0}^{t_k} e_A^0(t_k^+)P e_{\ominus A^*}^0(\sigma(\tau))F(\tau)\Delta_g\tau$$

$$- g\int_{t_k}^{\infty} e_A^0(t_k^+)P_1 e_{\ominus A^*}^0(\sigma(\tau))F(\tau)\Delta_g\tau$$

$$+ \sum_{t_j \leq t_k} e_A^0(t_k^+) P e_{\ominus A^*}^0(t_j^+) H_j -_g \sum_{t_j > t_k} e_A^0(t_k^+) P_1 e_{\ominus A^*}^0(t_j^+) H_j$$

$$= J_k \int_{t_0}^{t_k} e_A^0(t_k) P e_{\ominus A^*}^0(\sigma(\tau)) F(\tau) \Delta_g \tau$$

$$-_g J_k \int_{t_k}^{\infty} e_A^0(t_k) P_1 e_{\ominus A^*}^0(\sigma(\tau)) F(\tau) \Delta_g \tau$$

$$+ J_k \sum_{t_j < t_k} e_A^0(t_k) P e_{\ominus A^*}^0(t_j^+) H_j -_g J_k \sum_{t_j \geq t_k} e_A^0(t_k^+) P_1 e_{\ominus A^*}^0(t_j^+) H_j$$

$$+ e_A^0(t_k^+) P e_{\ominus A^*}^0(t_j^+) H_k + e_A^0(t_k^+) P_1 e_{\ominus A^*}^0(t_j^+) H_k$$

$$= J_k \mathscr{U}(F, \mathscr{H})(t_k) + H_k.$$

This shows that the second equation of (3.35) holds. The third equation of (3.35) hold straightforwardly. Conclusively, we obtain that $\mathscr{U}(F, \mathscr{H})$ is the desired solution.

Finally, under the assumption of the exponential dichotomy, the zero solution of the linear homogeneous SDE

$$\Delta_g U(t) = A(t) U(t)$$

is the unique solution that is bounded on $\mathbb{T}^+$. This is true because of Lemma A.2. Let U_1, U_2 both be the solutions of ISDE (3.35). From the first equation of (3.35), it follows that

$$\Delta_g U_1(t) -_g A(t) U_1(t) = \Delta_g U_2(t) -_g A(t) U_2(t)$$
$$= F(t) \quad \text{for } t \neq t_k.$$

By Proposition 2.4(d1), this yields

$$\Delta_g (U_1(t) -_g U_2(t)) = A(t)(U_1(t) -_g U_2(t)),$$

which further shows that $U_1 -_g U_2$ is a solution of

$$\Delta_g U(t) = A(t) U(t).$$

Hence,

$$U_1(t) -_g U_2(t) = \{0\},$$

that is,

$$U_1(t) = U_2(t) \quad \text{for } t \neq t_k.$$

Since U_i $(i = 1, 2)$ is left continuous at t_k, we have, for $k \in \mathbb{N}_0$,

$$U_1(t_k) = \lim_{t \to t_k^-} U_1(t) = \lim_{t \to t_k^-} U_2(t) = U_2(t_k).$$

This guarantees that $U_1 = U_2$ and the uniqueness of solutions follows. This completes the proof. □

Corollary 3.3. *Suppose $F \in BC[\mathbb{T}^+, K_c^n]$ and that the conditions of Theorem 3.17 hold. Then, ISDE*

$$\begin{cases} \Delta_g U(t) = A(t)U(t) + F(t, U), & t \neq t_k, \ k \in \mathbb{N}_0, \\ U_{t_k^+} = J_k U(t_k), & k \in \mathbb{N}_0, \\ U(t_0) = U_0 \end{cases} \tag{3.42}$$

has a unique bounded solution $\mathscr{U}(F, \mathscr{H})$ on $\mathbb{T}^+$ satisfying

$$\mathscr{U}(F, \mathscr{H})(t)$$

$$= \begin{cases} \mathscr{U}_0(t, t_0, U_0) & \text{for } t \in [0, t_0]_{\mathbb{T}^+} \\ \displaystyle\int_{t_0}^t e_A^0(t) P e_{\ominus A^*}^0(\sigma(\tau)) F(\tau) \Delta_g \tau \\ -g \displaystyle\int_t^\infty e_A^0(t) P_1 e_{\ominus A^*}^0(\sigma(\tau)) F(\tau) \Delta_g \tau & \text{for } t \in (t_0, \infty)_{\mathbb{T}^+}, \end{cases}$$

where $P_1 = (\mathcal{I} - P)$. Moreover, in the absence of the initial condition, ISDE (3.42), has a unique solution

$$\mathscr{U}(F, \mathscr{H})(t) = \int_{t_0}^t e_A^0(t) P e_{\ominus A^*}^0(\sigma(\tau)) F(\tau) \Delta_g \tau$$

$$-g \int_t^\infty e_A^0(t) P_1 e_{\ominus A^*}^0(\sigma(\tau)) F(\tau) \Delta_g \tau \quad \text{for } t \in \mathbb{T}^+. \tag{3.43}$$

Proof. This follows immediately by taking $H_k = 0$ $(k \in \mathbb{N}_0)$, that is, $\mathscr{H} = \Theta$, in Theorem 3.17. □

Remark 3.5. If we consider the complete metric space $BC_0 = BC_0[\mathbb{T}^+, K_c^n]$ consisting of functions $U \in BC$ satisfying

$$U_{t_k^+} = J_k(U(t_k)),$$

then, under the assumptions of Corollary 3.3, without the condition (H2), the conclusion of Corollary 3.3 is still valid.

Now, we are in a position to discuss the existence of solutions to nonlinear ISDE (3.26). We need the following well-known fixed point theorem which is the foundational tool to prove our main results.

Theorem 3.18. *Let $f\colon (K_c^n, D) \to (K_c^n, D)$ be continuous and compact mapping. Suppose the set*

$$\mathscr{M} = \{A \in K_c^n \colon \text{there exists a constant } \lambda \in (0,1) \text{ such that}$$
$$A = \lambda f(A)\} \tag{3.44}$$

is bounded, that is, there exists a positive constant c such that $\|A\| \le c$ for all $A \in \mathscr{M}$. Then, the operator f has a fixed point in $\mathscr{B} := \{A \in K_c^n \colon \|A\| \le c\}$.

Proof. Let $\mathscr{D} := \{A \in K_c^n \colon \|A\| < r\}$ with $r > 0$. Then, by Proposition A.1, there exists a compact, convex set $\mathscr{N} \subset K_c^n$ such that $\mathscr{D}$ is an open subset of $\mathscr{N}$ and $\theta \in \mathscr{D}$. Let $\partial\mathscr{D}$ be the boundary of $\mathscr{D}$. If f has a fixed point on $\partial\mathscr{D}$, then the conclusion is obvious. Suppose that f has no fixed point on $\partial\mathscr{D}$, and $B \ne \lambda f(B)$ for all $B \in \partial\mathscr{D}$ and $\lambda \in [0,1]$. Consider

$$\mathscr{E} = \{A \in \overline{\mathscr{D}} \colon A = t f(A) \text{ for some } t \in [0,1]\}.$$

Note that $\mathscr{E} \ne \emptyset$, since $\theta \in \mathscr{D}$. Further, the continuity of f implies that $\mathscr{E}$ is closed. Since $\mathscr{E} \cap \partial\mathscr{D} \ne \emptyset$, by Lemma A.1, there exists a continuous function $\eta\colon \overline{\mathscr{D}} \to [0,1]$ with $\eta(\mathscr{E}) = 1$ and $\eta(\partial\mathscr{D}) = 0$. Let

$$G(A) = \begin{cases} \eta(A)f(A) & \text{for } A \in \overline{\mathscr{D}} \\ \{0\} & \text{for } A \in \mathscr{N} \setminus \overline{\mathscr{D}}. \end{cases}$$

Since f is continuous and compact, and η is continuous, it follows immediately that $G\colon \mathscr{N} \to \mathscr{N}$ is continuous and compact map. Therefore, by Theorem A.3, G has a fixed point $A \in \mathscr{N}$. Note that $A \in \mathscr{E}$ and hence

$$A = G(A)$$
$$= \eta(A)f(A)$$
$$= f(A).$$

As a result, f has a fixed point in $\overline{\mathscr{D}}$. Now, consider the set

$$\mathscr{B}_k = \left\{ A \in K_c^n : \|A\| < c + \frac{1}{k} \right\}.$$

The above discussion guarantees that f has a fixed point $A_k \in \overline{\mathscr{B}_k}$ for $k \in \mathbb{N}$. In view of the compactness of f, there exists a subsequence $\{A_{k_i}\}$ of A_k such that $f(A_{k_i}) \to A$, this yields, $A_{k_i} \to A$. From the continuity of f, it follows that $A = f(A)$. Obviously, $A \in \mathscr{B}$. This completes the proof. $\qquad\square$

Theorem 3.19. *Assume that* (3.32) *admits an exponential dichotomy on* $(t_0, \infty)_{\mathbb{T}^+}$ *with positive constants* k *and* a, *a projection* P, *and the* $n \times n$-*matrix-valued function* A *which satisfy hypothesis* (H2). *Suppose* $F \in PC[\mathbb{T}^+ \times \mathcal{C}_{\mathrm{rd}}, K_c^n]$ *satisfies the following conditions.*

(i) *For each* $\Phi \in PC^1$, *the set-valued function* $F_\Phi(s) = F(s, \Phi(s))$ *satisfies the hypothesis* (H1).

(ii) *There exists function* $h : [0, t_0]_{\mathbb{T}^+} \to \mathbb{R}^+$ *with*

$$\sup_{t \in [0, t_0]_{\mathbb{T}^+}} \int_0^t |e_A(t, s)| h(s) \Delta s < 1$$

such that

$$D[F(t, U), F(t, V)] \leq h(t) D[U, V]$$

for all $t \in [0, t_0]_{\mathbb{T}^+}$ *and* $U, V \in PC^1$.

(iii) *There exists function* $\psi : (t_0, \infty)_{\mathbb{T}^+} \to \mathbb{R}^+$ *which is delta integrable on* $(t_0, \infty)_{\mathbb{T}^+}$ *such that*

$$\|F(t, \Phi(t))\| \leq \psi(t)$$

for each $(t, \Phi) \in (t_0, \infty)_{\mathbb{T}^+} \times BC$ *and* $t \neq t_k$, $k \in \mathbb{N}_0$.

Then, the nonlinear ISDE

$$\begin{cases} \Delta_g U(t) = A(t) U(t) + F(t, U), & t \neq t_k, \ k \in \mathbb{N}_0, \\ U_{t_k^+} = J_k U(t_k) + H_k, & k \in \mathbb{N}_0, \\ U(t_0) = U_0 \end{cases} \qquad (3.45)$$

has a bounded solution on $\mathbb{T}^+$ *given by*

$$U(t) = \begin{cases} e_A^0(t)U_0 + \displaystyle\int_{t_0}^t e_A(t,\sigma(s))F(s,U)\Delta_g s \quad for\ t \in [0,t_0]_{\mathbb{T}^+} \\[2mm] \displaystyle\int_{t_0}^t e_A^0(t)Pe_{\ominus A^*}^0(\sigma(s))F(s)\Delta_g s + \sum_{t_j < t} e_A^0(t_k)Pe_{\ominus A^*}^0(t_j^+)H_j \\[2mm] -g\displaystyle\int_t^\infty e_A^0(t)P_1 e_{\ominus A^*}^0(\sigma(s))F(s)\Delta_g s \\[2mm] -g\displaystyle\sum_{t_j \geq t} e_A^0(t_k^+)P_1 e_{\ominus A^*}^0(t_j^+)H_j \quad for\ t \in (t_0,\infty)_{\mathbb{T}^+}. \end{cases}$$

$$(3.46)$$

Proof. For fixed $\Phi \in PC^1$, consider ISDE (3.45) with $F(t,\Phi) = F_\Phi(t)$ instead of $F(t,U)$. From Theorem 3.17, it follows that the corresponding linear ISDE has a unique solution $\mathscr{U}(F_\Phi, \mathscr{H})$ such that

$$\mathscr{U}(F_\Phi, \mathscr{H}) = \delta_1(t)\mathscr{U}_0(t,\Phi) + \delta_2(t)\mathscr{U}_1(t,\Phi),$$

where

$$\mathscr{U}_0(t,\Phi) = e_A^0(t)U_0 + \int_{t_0}^t e_A(t,\sigma(\tau))F_\Phi(\tau)\Delta_g\tau,$$

$$\mathscr{U}_1(t) = \int_{t_0}^t e_A^0(t)Pe_{\ominus A^*}^0(\sigma(\tau))F_\Phi(\tau)\Delta_g\tau$$

$$-g\int_t^\infty e_A^0(t)P_1 e_{\ominus A^*}^0(\sigma(\tau))F_\Phi(\tau)\Delta_g\tau$$

$$+\sum_{t_j < t} e_A^C(t)Pe_{\ominus A^*}^0(t_j^+)H_j - g\sum_{t_j \geq t} e_A^0(t)Pe_{\ominus A^*}^0(t_j^+)H_j,$$

and

$$\delta_1(t) = \begin{cases} 1 & for\ t \in [0,t_0]_{\mathbb{T}^+} \\ 0 & if\ t \in (t_0,\infty)_{\mathbb{T}^+}, \end{cases}$$

and

$$\delta_2(t) = \begin{cases} 0 & for\ t \in [0,t_0]_{\mathbb{T}^+} \\ 1 & for\ t \in (t_0,\infty)_{\mathbb{T}^+}. \end{cases}$$

Let

$$f(\Phi, \Psi) = f_1[\Phi] + f_2[\Psi]$$

with

$$f_1[\Phi](t) = \delta_1(t)\mathscr{U}_1(t, \Phi)$$

and

$$f_2[\Psi](t) = \delta_2(t)\mathscr{U}_1(t, \Psi).$$

Then, the mapping $f \colon PC^1 \times PC^1 \to PC^1$ and fixed point of f is a solution of nonlinear ISDE (3.45). To this end, we have to prove that f_1 and f_2 both have a fixed point in PC^1. We first observe that condition (ii) guarantees f_1 to be a contractive mapping, and therefore, by the Banach fixed point theorem, f_1 has a unique fixed point $\widehat{\Phi}_1 \in PC^1[[0, t_0]_{\mathbb{T}+}, K_c^n]$. Let $\widehat{\Phi}$ be a set-valued function such that

$$\widehat{\Phi}(t) = \begin{cases} \widehat{\Phi}_1(t) & \text{for } t \in [0, t_0]_{\mathbb{T}+} \\ \{0\} & \text{for } t \in (t_0, \infty)_{\mathbb{T}+}. \end{cases}$$

Then, we see that $\widehat{\Phi}$ is a unique fixed point of f_1 in PC^1. Next, we prove that f_2 has a fixed point in PC^1. As an analogous of the arguments of the proof of Theorem 3.17, we see that f_2 is bounded and continuous on PC^1. We shall verify that f_2 is equicontinuous. In fact, for $t_1, t_2 \in [0, t_0]_{\mathbb{T}+}$ with $t_1 < t_2$ and $\Phi \in PC^1$, in virtue of the properties of the Hausdorff distance and Proposition 2.5, combining our hypotheses, we have

$$D\left[f_2[\Psi](t_2), f_2[\Psi](t_1)\right]$$

$$= D\left[\int_{t_0}^{t_2} e_A^0(t_2) P e_{\ominus A^*}^0(\sigma(s)) F_\Psi(s) \Delta_g s \right.$$

$$-g \int_{t_2}^{\infty} e_A(t_2) P_1 e_{\ominus A^*}^0(\sigma(s)) F_\Psi(s) \Delta_g s$$

$$+ \sum_{t_j < t_2} e_A^0(t_2) P e_{\ominus A^*}^0(t_j^+) H_j -g \sum_{t_j \geq t_2} e_A^0(t_2) P_1 e_{\ominus A^*}^0(t_j^+) H_j,$$

$$\int_{t_0}^{t_1} e_A^0(t_1) P e_{\ominus A^*}^0(\sigma(s)) F_\Psi(s) \Delta_g s$$

$$-g \int_{t_1}^{\infty} e_A^0(t_1) P_1 e_{\ominus A^*}^0(\sigma(s)) F_\Psi(s) \Delta_g s$$

$$+ \sum_{t_j < t_1} e_A^0(t_1) P e_{\ominus A^*}^0(t_j^+) H_j -g \sum_{t_j \geq t_1} e_A^0(t_1) P_1 e_{\ominus A^*}^0(t_j^+) H_j \Bigg]$$

$$= D \Bigg[\int_{t_0}^{t_1} e_A^0(t_2) P e_{\ominus A^*}^0(\sigma(s)) F_\Psi(s) \Delta_g s$$

$$+ \int_{t_1}^{t_2} e_A^0(t_2) P e_{\ominus A^*}^0(\sigma(s)) F_\Psi(s) \Delta_g s$$

$$-g \int_{t_2}^{\infty} e_A(t_2) P_1 e_{\ominus A^*}^0(\sigma(s)) F_\Phi(s) \Delta s$$

$$+ \sum_{t_j < t_2} e_A^0(t_2) P e_{\ominus A^*}^0(t_j^+) H_j -g \sum_{t_j \geq t_2} e_A^0(t_2) P_1 e_{\ominus A^*}^0(t_j^+) H_j,$$

$$\int_{t_0}^{t_1} e_A^0(t_1) P e_{\ominus A^*}^0(\sigma(s)) F_\Phi(s) \Delta_g s$$

$$-g \int_{t_1}^{t_2} e_A^0(t_1) P_1 e_{\ominus A^*}^0(\sigma(s)) F_\Phi(s) \Delta_g s$$

$$-g \int_{t_2}^{\infty} e_A(t_1) P_1 e_{\ominus A^*}^0(\sigma(s)) F_\Phi(s) \Delta_g s$$

$$+ \sum_{t_j < t_1} e_A^0(t_1) P e_{\ominus A^*}^0(t_j^+) H_j -g \sum_{t_j \geq t_1} e_A^0(t_1) P_1 e_{\ominus A^*}^0(t_j^+) H_j \Bigg]$$

$$\leq \mathscr{D}_1 + \mathscr{D}_2 + \mathscr{D}_3 + \mathscr{D}_4,$$

where

$$\mathscr{D}_1 = D \Bigg[\int_{t_0}^{t_2} e_A^0(t_2) P e_{\ominus A^*}^0(\sigma(s)) F_\Psi(s) \Delta_g s,$$

$$\int_{t_0}^{t_1} e_A^0(t_1) P e_{\ominus A^*}^0(\sigma(s)) F_\Psi(s) \Delta_g s \Bigg]$$

$$\leq \int_{t_0}^{t_2} D\left[e_A^0(t_2)Pe_{\ominus A^*}^0(\sigma(s))F_\Psi(s), e_A^0(t_1)Pe_{\ominus A^*}^0(\sigma(s))F_\Psi(s)\right]\Delta s,$$

$$\mathscr{D}_2 = D\left[\int_{t_1}^{t_2} e_A^0(t_2)Pe_{\ominus A^*}^0(\sigma(s))F_\Psi(s)\Delta_g s,\right.$$

$$\left. -_g\int_{t_0}^{t_1} e_A^0(t_1)Pe_{\ominus A^*}^0(\sigma(s))F_\Psi(s)\Delta_g s\right]$$

$$\leq \int_{t_1}^{t_2} D\left[\left|e_A^0(t_2)Pe_{\ominus A^*}^0(\sigma(s))\right|\right.$$

$$\left. + \left|e_A^0(t_1)P_1 e_{\ominus A^*}^0(\sigma(s))\right|\right]\|F_\Psi(s)\|\Delta s,$$

$$\mathscr{D}_3 = D\left[\int_{t_2}^{\infty} e_A^0(t_2)P_1 e_{\ominus A^*}^0(\sigma(s))F_\Psi(s)\Delta_g s,\right.$$

$$\left.\int_{t_2}^{\infty} e_A^0(t_1)P_1 e_{\ominus A^*}^0(\sigma(s))F_\Psi(s)\Delta_g s\right]$$

$$\leq \int_{t_2}^{\infty} D\left[e_A^0(t_2)P_1 e_{\ominus A^*}^0(\sigma(s))F_\Psi(s), e_A^0(t_1)P_1 e_{\ominus A^*}^0(\sigma(s))F_\Psi(s)\right]\Delta s,$$

and

$$\mathscr{D}_4 = D\left[\sum_{t_j<t_2} e_A^0(t_2)Pe_{\ominus A^*}^0(t_j^+)H_j -_g \sum_{t_j\geq t_2} e_A^0(t_2)P_1 e_{\ominus A^*}^0(t_j^+)H_j,\right.$$

$$\left.\sum_{t_j<t_1} e_A^0(t_1)Pe_{\ominus A^*}^0(t_j^+)H_j -_g \sum_{t_j\geq t_1} e_A^0(t_1)P_1 e_{\ominus A^*}^0(t_j^+)H_j\right].$$

Letting $t_2 \to t_1$, and from condition (iii), together with the continuity of $e_A^0(t)$ and Δ-integrability of ψ, it follows that f_2 is equicontinuous with respect to $\Psi \in PC^1$. In virtue of Theorem A.2, we obtain that f_2 is a continuous compact operator on PC^1. Now, we prove that the set $\mathscr{M}$ defined by (3.44) is bounded. If it is not bounded, then there exist $(\lambda_m, U_m) \in (0,1) \times PC^1$ such that

$$U_m = \lambda_m f_2(U_m) \quad \text{and} \quad \|U_m\|_0 > m \quad \text{for } m \in \mathbb{N}.$$

On the other hand, for $t \in (t_0, \infty)_{\mathbb{T}^+}$, from (3.37) and (3.40), Definition 3.10, and assumption (iii), we have

$$\|U_m(t)\| \leq \lambda_m \int_{t_0}^t \left\| e_A^0(t) P e_{\ominus A^*}^0(\sigma(s)) F_{U_m}(s) \right\| \Delta s$$

$$+ \lambda_m \int_t^\infty \left\| e_A^0(t) P_1 e_{\ominus A^*}^0(\sigma(s)) F_{U_m}(s) \right\| \Delta s$$

$$+ \lambda_m \left\| \sum_{t_j < t} e_A^0(t) P e_{\ominus A^*}^0(t_j^+) H_j \right\|$$

$$+ \lambda_m \left\| \sum_{t_j \geq t} e_A^0(t, 0) P_1 e_{\ominus A^*}^0(t_j^+) H_j \right\|$$

$$\leq k \int_0^t e_{\ominus a}(t, \sigma(s)) \, \|F_{U_m}(s)\| \, \Delta s$$

$$+ k \int_t^\infty e_{\ominus_2}(\sigma(s), t) \, \|F_{U_m}(s)\| \, \Delta s$$

$$+ \frac{k}{1-\chi} \|\mathscr{H}\|_\infty + \frac{k e_a(\sigma(0), \sigma(t_1))}{1-\chi} \|\mathscr{H}\|_\infty$$

$$\leq k \int_0^t e_{\ominus a}(t, \sigma(s)) \psi(s) \Delta s + k \int_t^\infty e_{\ominus a}(\sigma(s), t) \psi(s) \Delta s$$

$$+ \frac{k}{1-\chi} \|\mathscr{H}\|_\infty + \frac{k e_a(\sigma(0), \sigma(t_1))}{1-\chi} \|\mathscr{H}\|_\infty$$

$$\leq 2k \int_0^\infty \psi(s) \Delta s + \frac{k}{1-\chi} \|\mathscr{H}\|_\infty + \frac{k e_a(\sigma(0), \sigma(t_1))}{1-\chi} \|\mathscr{H}\|_\infty.$$

In view of the arbitrariness of $t \in (t_0, \infty)_{\mathbb{T}^+}$, we obtain

$$\|U_m\|_0 \leq 2k \int_0^\infty \psi(s) \Delta s + \frac{k}{1-\chi} \|\mathscr{H}\|_\infty + \frac{k e_a(\sigma(0), \sigma(t_1))}{1-\chi} \|\mathscr{H}\|_\infty.$$

Since m is arbitrary, take it to be so large that

$$m > 2k \int_0^\infty \psi(s) \Delta s + \frac{k}{1-\chi} \|\mathscr{H}\|_\infty + \frac{k e_a(\sigma(0), \sigma(t_1))}{1-\chi} \|\mathscr{H}\|_\infty,$$

and thus, we get a contradiction. Consequently, the set $\mathscr{M}$ is bounded. Now, Theorem 3.18 guarantees that f_2 has a fixed point

$\widehat{\Psi}_1(t) \in PC^1[(t_0, \infty)_{\mathbb{T}+}, K_c^n]$. Let $\widehat{\Psi}$ be a set-valued function such that

$$\widehat{\Psi}(t) = \begin{cases} \widehat{\Psi}_1(t) & \text{for } t \in (t_0, \infty)_{\mathbb{T}+} \\ \{0\} & \text{for } t \in [0, t_0]_{\mathbb{T}+}. \end{cases}$$

Then, $\widehat{\Psi}(t)$ is a fixed point of f_2 in PC^1. Finally, we prove that f has a fixed point in $PC^1 \times PC^1$. Let $\Psi \in PC^1$ be fixed and define the mapping

$$G[\Psi] = f(\Phi, \Psi) = f_1[\Phi] + f_2[\Psi].$$

Then, from the fact that f_1 has a unique fixed point $\widehat{\Phi}$ satisfying $\widehat{\Phi}(t) = \{0\}$ for $t \in (t_0, \infty)_{\mathbb{T}+}$ and $f_2[\Psi](t) = \{0\}$ for $t \in [0, t_0]_{\mathbb{T}+}$, it follows that

$$G[\Psi] = f_1[\widehat{\Phi}] + f_2[\Psi] = \widehat{\Phi} + f_2[\Phi].$$

Similarly, we see that $G\colon PC^1 \to PC^1$ satisfies all conditions of Theorem 3.18. Therefore, G has a fixed point $\Gamma \in PC^1$ satisfying

$$\Gamma(t) = \widehat{\Phi}(t) \quad \text{for } t \in [0, t_0]_{\mathbb{T}+}.$$

This further implies that

$$\Gamma = f(\widehat{\Phi}, \Gamma) = f(\Gamma, \Gamma).$$

That is, f has a fixed point in PC^1. This completes the proof. $\square$

Corollary 3.4. *Assume that all conditions of Theorem 3.19 are satisfied, except for* (H2). *In addition, instead of* PC^1, *we consider the complete metric space* $PC^0[\mathbb{T}^+, K_c^n]$ *consisting of the functions* $U \in BC_0$ *and their* Δ_g-*derivatives exist. Then, the nonlinear ISDE* (3.26) *has at least bounded solution.*

3.5 Almost Periodicity

In this section, we first propose the concept of almost periodicity of set-valued functions on almost periodic time scales. Then based on the exponential dichotomy, we discuss the existence and uniqueness of almost periodic solutions of SDEs on time scales. As an application, several examples are presented to exhibit that one can obtain the existence of almost periodic solutions to interval-valued dynamic equations and their discrete counterparts in an united means.

3.5.1 *Almost periodicity of set-valued functions*

A subset S of $\mathbb{T}$ is called relatively dense provided there exists a positive number l such that $[t, t+l] \cap S \neq \emptyset$ for all $t \in \mathbb{T}$. Below, we define the concept of almost periodic time scale and almost periodic function.

Definition 3.11. Let $\mathcal{D}$ be a collection of nonempty subsets of $\mathbb{R}$. A time scale $\mathbb{T}$ is called almost periodic with respect to $\mathcal{D}$, provided the set

$$C^* = \left\{ \omega \in \bigcap_{\mathbb{D} \in \mathcal{D}} \mathbb{D} : t \pm \omega \in \mathbb{T} \text{ for each } t \in \mathbb{T} \right\} \neq \{0\}$$

is nonempty. The set C^* is called the smallest almost periodic subset of $\mathbb{T}$.

It is clear that Definition 3.11 includes the concept of periodic time scales given in Bi *et al.* (2008) when we take $\mathcal{D} = \{\mathbb{R}\}$. Throughout this section, we restrict our discussion on almost periodic time scales.

Definition 3.12. Let $\mathbb{T}$ be an almost periodic time scale with respect to $\mathcal{D}$. A set-valued function $F \in \mathcal{C}_{\mathrm{rd}}(\mathbb{T}, K_c^n)$ is called almost periodic provided for any given $\varepsilon > 0$, the set

$$\mathscr{E}(\varepsilon, F) = \{\omega \in C^* : D[F(t+\omega), F(t)] < \varepsilon \quad \text{for all } t \in \mathbb{T}\} \quad (3.47)$$

is relatively dense in $\mathbb{T}$, that is, there exists an $l = l(\varepsilon) > 0$ such that each interval of length l contains at least one $\omega = \omega(\varepsilon) \in \mathscr{E}(\varepsilon, F)$, i.e., $\omega \in C^*$ and

$$D[F(t+\omega), F(t)] < \varepsilon \quad \text{for all } t \in \mathbb{T}.$$

The set $\mathscr{E}(\varepsilon, F)$ is called the ε-almost periodic set of F, ω is called an ε-almost period of F and l is called the inclusion length of $\mathscr{E}(\varepsilon, F)$.

A function $F \in \mathcal{C}_{\mathrm{rd}}(\mathbb{T} \times K_c^n, K_c^n)$ is said to be almost periodic in $t \in \mathbb{T}$ uniformly for $U \in K_c^n$, provided for any given $\varepsilon > 0$, the set

$$\mathscr{E}(\varepsilon, F, \mathscr{S}) = \{\omega \in C^* : D[F(t+\omega, U), F(t, U)] < \varepsilon$$
$$\text{for all } (t, U) \in \mathbb{T} \times \mathscr{S}\}$$

is relatively dense in $\mathbb{T}$ for each compact subset $\mathscr{S}$ of K_c^n.

Note 3.1. Henceforth, we use the following notations.

(1) The set $AP(\mathbb{T}, K_c^n)$ denotes the collection consisting of all almost periodic set-valued functions from $\mathbb{T}$ into K_c^n.
(2) The set $AP_u(\mathbb{T} \times K_c^n, K_c^n)$ denotes the collection consisting of all almost periodic set-valued functions from $\mathbb{T} \times K_c^n$ into K_c^n in $t \in \mathbb{T}$ uniformly for $U \in K_c^n$.
(3) Let $\alpha = \{\alpha_m\}_{m \in \mathbb{N}}$ be a sequence. We write $T_\alpha F(t, U) = G(t, U)$, provided

$$\lim_{m \to \infty} F(t + \alpha_m, U) = G(t, U)$$

exists for $F \in \mathcal{C}_{\mathrm{rd}}(\mathbb{T} \times K_c^n, K_c^n)$.
(4) The hull of F is the set, $H(F)$, consists of the functions $G \colon \mathbb{T} \times K_c^n \to K_c^n$ for which $T_\alpha F(t, U) = G(t, U)$ uniformly holds on $\mathbb{T} \times K_c^n$ for some sequence $\alpha \subset \mathcal{C}^*$.

Remark 3.6. If $F \in \mathcal{C}(\mathbb{T} \times K_c^n, K_c^n)$ is almost periodic in t uniformly for $U \in K_c^n$, then F is uniformly continuous and bounded on $\mathbb{T} \times K_c^n$.

Theorem 3.20. *Let $F \in \mathcal{C}(\mathbb{T} \times K_c^n, K_c^n)$. Then, $F(t, U)$ be almost periodic in $t \in \mathbb{T}$ uniformly for $U \in K_c^n$ if and only if, for any given sequence $\alpha' \subset \mathcal{C}^*$, there exist a subsequence $\alpha \subset \alpha'$ and $G \in \mathcal{C}_{\mathrm{rd}}(\mathbb{T} \times K_c^n, K_c^n)$ such that*

$$T_\alpha F(t, U) = G(t, U)$$

uniformly on $\mathbb{T} \times \mathscr{S}$ for any compact subset $\mathscr{S}$ of K_c^n and G is also almost periodic in t uniformly for $U \in K_c^n$.

Proof. For any $\varepsilon > 0$ and $\mathscr{S} \subset K_c^n$, let $l = l\left(\frac{\varepsilon}{4}, \mathscr{S}\right)$ be an inclusion length of $\mathscr{E}(\varepsilon, F, \mathscr{S})$. For any given sequence $\alpha'_n \subset \mathcal{C}^*$, we denote $\alpha'_n = \beta'_n + \gamma'_n$, where $\beta'_n \in \mathscr{E}(\varepsilon, F, \mathscr{S})$, $\gamma'_n \in \mathcal{C}^*$, and $0 \leq \gamma'_n \leq 1$, $n \in \mathbb{N}$ (In fact, for any interval of length l, there exists $\beta'_n \in \mathscr{E}(\varepsilon, F, \mathscr{S})$, thus, we can choose a proper interval of length l such that $0 \leq \alpha'_n - \beta'_n \leq l$, and from the definition of $\alpha' \in \mathcal{C}^*$, it is easy to see that $\gamma'_n = \alpha'_n - \beta'_n$ and $\gamma'_n \in \mathcal{C}^*$). Therefore, there exists a subsequence $\gamma_n \subset \gamma'_n$ such that $\gamma_n \to s$ as $n \to \infty$, $0 \leq s \leq l$.

Suppose that F is almost periodic in t uniformly for $U \in K_c^n$. Then, by Remark 3.6, $F(t, U)$ is uniformly continuous on $\mathbb{T} \times K_c^n$. Hence, there exists $\delta(\varepsilon, U) > 0$ so that $|t_1 - t_2| < \delta$ implies

$$D[F(t_1, U), F(t_2, U)] < \frac{\varepsilon}{2} \quad \text{for } U \in K_c^n. \tag{3.48}$$

Since γ_n is a convergent sequence, there exists $N = N(\delta)$ so that $p, m \geq N$ implies $|\gamma_p - \gamma_m| < \delta$. Now, take $\alpha_n \subset \alpha'_n$, $\beta_n \subset \beta'_n$ such that $\alpha_n + \beta_n = \gamma_n$. Then, for $p, m \geq N$, we have

$$D[F(t + \beta_p - \beta_m, U), F(t, U)] \leq D[F(t + \beta_p - \beta_m, U), F(t + \beta_p, U)]$$
$$+ D[F(t + \beta_p, U), F(t, U)]$$
$$< \frac{\varepsilon}{4} + \frac{\varepsilon}{4}$$
$$= \frac{\varepsilon}{2},$$

that is,

$$(\alpha_p - \alpha_m) - (\gamma_p - \gamma_m) = (\beta_p - \beta_m) \in \mathscr{E}(\varepsilon, F, \mathscr{S}).$$

Hence, we can obtain

$$D[F(t + \alpha_p, U), F(t - \alpha_m, U)]$$
$$\leq \sup_{(t,U) \in \mathbb{T} \times \mathscr{S}} D[F(t + \alpha_p, U), F(t - \alpha_m, U)]$$
$$\leq \sup_{(t,U) \in \mathbb{T} \times \mathscr{S}} D[F(t + \alpha_p - \alpha_m, U), F(t, U)]$$
$$\leq \sup_{(t,U) \in \mathbb{T} \times \mathscr{S}} D[F(t + \alpha_p - \alpha_m, U), F(t + \gamma_p - \gamma_m, U)]$$
$$+ \sup_{(t,U) \in \mathbb{T} \times \mathscr{S}} D[F(t + \gamma_p - \gamma_m, U), F(t, U)]$$
$$< \frac{\varepsilon}{2} + \frac{\varepsilon}{2}$$
$$= \varepsilon.$$

Thus, we can take sequences $\alpha_n^{(k)}$, $k \in \mathbb{N}$, with $\alpha_n^{(k+1)} \subset \alpha_n^{(k)} \subset \alpha_n$ such that, for any $m, p \in \mathbb{N}$ and all $(t, U) \in \mathbb{T} \times K_c^n$, the following holds:

$$D[F(t + \alpha_p^{(k)}, U), F(t + \alpha_m^{(k)}, U)] < \frac{1}{k}, \quad k \in \mathbb{N}.$$

For all sequences $\alpha^{(k)}$, $k \in \mathbb{N}$, we can take a sequence $\beta_n = \alpha_n^{(n)}$ then, it is easy to see that

$$\{F(t + \beta_n, U)\} \subset \{F(t + \alpha_n, U)\}$$

for any $p, m \in \mathbb{N}$ with $p < m$, and all $(t, U) \in \mathbb{T} \times K_c^n$, the following holds:

$$D[F(t + \beta_p, U), F(t + \beta_m, U)] < \frac{1}{p}.$$

Therefore, $\{F(t + \beta_n, U)\}$ converges uniformly on $\mathbb{T} \times K_c^n$, that is,

$$T_\beta F(t, U) = G(t, U)$$

holds uniformly on $\mathbb{T} \times K_c^n$, where $\beta = \{\beta_n\} \subset \alpha$.

Next, we shall prove that $G \in \mathcal{C}_{\mathrm{rd}}(\mathbb{T} \times K_c^n, K_c^n)$. If this is not true, then there must exist $(t_0, U_0) \in \mathbb{T} \times K_c^n$ with right-dense t_0 such that $G(t, U)$ is not rd-continuous at this point t_0. Then there exist $\varepsilon_0 > 0$ and sequences $\{\delta_m\}$, $\{t_m\}$, and $\{U_m\}$, where $\delta_m > 0$, $\delta_m \to 0$ as $m \to \infty$, such that

$$|t_0 - t_m| + |U_0 - U_m| < \delta_m$$

and

$$D[G(t_0, U_0), G(t_m, U_m)] \geq \varepsilon_0. \tag{3.49}$$

Let $U = \{U_m\} \cup \{U_0\}$. Then obviously U is a compact subset of K_c^n. Hence, there exists $N = N(\varepsilon_0, U) \in \mathbb{N}$ so that $m > N$ implies

$$D[F(t_m + \beta_m, U_m) - G(t_m, U_m)] < \frac{\varepsilon_0}{3} \quad \text{for } m \in \mathbb{N},$$

$$D[F(t_0 + \beta_n, U_0), G(t_0, U_0)] < \frac{\varepsilon_0}{3}. \tag{3.50}$$

By virtue of the uniform continuity of $F(t, U)$ on $\mathbb{T} \times K_c^n$, for sufficiently large m, we have

$$D[F(t_0 + \beta_n, U_0), F(t_m + \beta_m, U_m)] < \frac{\varepsilon}{3}. \tag{3.51}$$

From (3.50) and (3.51), we get

$$D[G(t_0, U_0), G(t_m, U_m)] < \varepsilon_0, \tag{3.52}$$

this contradicts to (3.49). Thus, $G(t, U)$ is rd-continuous on $\mathbb{T} \times K_c^n$.

Finally, for any compact set $\mathscr{S} \subset K_c^n$ and given $\varepsilon > 0$, one can take $\tau \in \mathcal{E}(\varepsilon, F, \mathscr{S})$ such that for all $(t, U) \in \mathbb{T} \times \mathscr{S}$, the following holds:

$$D[F(t + \beta_n + \tau, U), F(t + \beta_n, U)] < \varepsilon. \tag{3.53}$$

Let $n \to \infty$. Then, for all $(t, U) \in \mathbb{T} \times K_c^n$, we have

$$D[G(t + \tau, U), G(t, U)] \leq \varepsilon, \qquad (3.54)$$

which implies that $\mathcal{E}(\varepsilon, G, \mathscr{S})$ is relatively dense. Therefore, $G(t, U)$ is almost periodic in t uniformly for $U \in K_c^n$.

Conversely, suppose that any given sequence $\alpha' \subset C^*$, there exist a subsequence $\alpha \subset \alpha'$ and $G \in \mathcal{C}_{\mathrm{rd}}(\mathbb{T} \times K_c^n, K_c^n)$ such that

$$T_\alpha F(t, U) = G(t, U)$$

uniformly on $\mathbb{T} \times \mathscr{S}$ for a compact subset $\mathscr{S}$ of K_c^n and G is almost periodic in t uniformly for $U \in K_c^n$. To prove that F is almost periodic in $t \in \mathbb{T}$ uniformly for $U \in K_c^n$. For contradiction, if this is not true, then there exist $\varepsilon > 0$ and $\mathscr{S} \subset K_c^n$ such that, for any sufficiently large $l > 0$, we can find an interval of length l and there is no ε_0-translation numbers of $F(t, U)$ in this interval, that is, every point in this interval is not in $\mathcal{E}(\varepsilon, F, \mathscr{S})$.

We can take a number $\alpha_1' \in C^*$ and find an interval (a_1, b_1) with $b_1 - a_1 > 2|\alpha_1'|$, where $a_1, b_1 \in C^*$ such that there is no ε_0-translation numbers of $F(t, U)$ in (a_1, b_1). Next, taking

$$\alpha_2' = \frac{1}{2}(a_1 + b_1),$$

we have $\alpha_2' - \alpha_1' \in (a_1, b_1)$, so

$$\alpha_2' - \alpha_1' \notin \mathcal{E}(\varepsilon, F, \mathscr{S}).$$

Then, we can find an interval (a_2, b_2) with

$$b_2 - a_2 > 2(|\alpha_1'| + |\alpha_2'|),$$

where $a_2, b_2 \in C^*$ such that there is no ε_0-translation numbers of $F(t, U)$ in (a_2, b_2). Next, taking

$$\alpha_2' = \frac{1}{2}(a_2 + b_2),$$

obviously,

$$\alpha_3' - \alpha_2', \alpha_3' - \alpha_1' \notin \mathcal{E}(\varepsilon, F, \mathscr{S}).$$

Repeating these processes we can find $\alpha_4', \alpha_5', \ldots$, such that

$$\alpha_i' - \alpha_j' \notin \mathcal{E}(\varepsilon, F, \mathscr{S}) \quad \text{for } i > j.$$

Hence, for any $i \neq j$, $i, j \in \mathbb{N}$ (without any restriction, let $i > j$) and for $U \in \mathscr{S}$, we have

$$\sup_{(t,U) \in \mathbb{T} \times \mathscr{S}} D[F(t + \alpha_i', U), F(t + \alpha_j', U)]$$

$$= \sup_{(t,U) \in \mathbb{T} \times \mathscr{S}} D[F(t + \alpha_i' - \alpha_j', U), F(t, U)]$$

$$\geq \varepsilon.$$

Thus, there is no uniformly convergent subsequence of

$$\{F(t_{\alpha_n'}, x)\} \quad \text{for } (t, U) \in \mathbb{T} \times \mathscr{S},$$

which is contradiction. Hence, $F(t, U)$ is almost periodic in t uniformly for $U \in K_c^n$. This completes the proof. $\qquad \square$

Remark 3.7. Another necessary and sufficient condition for the almost periodicity of $F(t, U)$ in $t \in \mathbb{T}$ uniformly for $U \in K_c^n$ is that, for each pair of sequences $\alpha', \beta' \subset C^*$, there exist common subsequences $\alpha \subset \alpha'$, $\beta \subset \beta'$ such that

$$T_{\alpha+\beta} F(t, U) = T_\alpha T_\beta F(t, U).$$

If $F: \mathbb{T} \to \mathbb{R}$ is a single-valued function and $C^* = \mathbb{R}$, then the almost periodicity of F is coincident with the almost periodicity, which given by Li and Wang in Li and Wang (2011, 2012), and for $\mathbb{T} = \mathbb{R}$, $F \in AP(\mathbb{T}, K_c^n)$ if and only if F is almost periodic in the sense of the definition in Corduneanu (1989). Also, for $C^* = \mathbb{T} = \mathbb{Z}$, $F \in AP(\mathbb{T}, K_c^n)$ is coincident with the almost periodicity in the sense of the definition given by Cheban and Mammana in Cheban and Mammana (2004). Thus, we extend and unify the corresponding concept defined in Cheban and Mammana (2004), Corduneanu (1989) and Li and Wang (2011, 2012). Moreover, an ω-periodic function is indeed almost periodic if $\mathbb{T}$ is ω-periodic, this is discussed in Hong and Peng (2016). In addition, the following properties of almost periodic set-valued functions extend corresponding results of single-valued real functions.

Proposition 3.1. *Let $F, G \in AP(\mathbb{T}, K_c^n)$. Then,*

(a1) *If F is regulated, then F is bounded.*
(a2) *If F is rd-continuous (continuous), then F is uniformly rd-continuous (continuous) on $\mathbb{T}$.*

(a3) *The functions aF, $a \in \mathbb{R}$, $F \pm_g G$ and FG are almost periodic.*

(a4) *Let R_f denote the range of values of the single-valued function $f \in AP(\mathbb{T}, \mathbb{T})$ and the function $F \colon \mathbb{T} \to K_c^n$ uniformly continuous on R_f. Then the mapping $F \circ f \colon \mathbb{T} \to K_c^n$, defined as*

$$(F \circ f)(t) = F(f(t))$$

for each $t \in \mathbb{T}$, is almost periodic.

(a5) *If the sequence $\{F_m\}$ in $AP(\mathbb{T}, K_c^n)$ uniformly converges on $\mathbb{T}$ to F, then $F \in AP(\mathbb{T}, K_c^n)$.*

(a6) *If $F, G \in AP_u(\mathbb{T} \times K_c^n, K_c^n)$, then, for any $\varepsilon > 0$,*

$$\mathscr{E}(F, \varepsilon, \mathscr{S}) \cap \mathscr{E}(G, \varepsilon, \mathscr{S})$$

is a nonempty relatively dense set in $\mathbb{T}$ for each compact subset $\mathscr{S}$ of K_c^n.

Proof. We check only (a1) and other proofs are left as an exercise, which are similar to those discussed in Corduneanu (1989) and Li and Wang (2011). Let $\mathbb{T}_0 = \mathbb{T} \cup \{0\}$. If $0 \notin \mathbb{T}$, then 0 is isolated in $\mathbb{T}_0$. Hence, F is still regulated on $\mathbb{T}_0$. Without any restriction, we assume that $\mathbb{T}_0 = \mathbb{T}$.

First, suppose that $\mathbb{T}$ is symmetrical about zero and take $\varepsilon = 1$. By means of the Lemma 2.2, F is bounded on the interval $[0, l(1)]_\mathbb{T}$, where $l(1)$ is the inclusion length of $\mathscr{E}(1, F)$. Hence, we can assume that

$$M = \max_{t \in [0, l(1)]_\mathbb{T}} \|F(t)\|.$$

From Definition 3.12, it follows that there exists $s \in [-t, -t + l(1)]_\mathbb{T}$ such that $s + t \in [0, l(1)]_\mathbb{T}$ for any $t \in \mathbb{T}$, and

$$\|F(s + t)\| \leq M$$

and

$$D[F(s + t), F(t)] < 1.$$

Consequently,

$$\|F(t)\| \leq M + 1 \quad \text{for each } t \in \mathbb{T}.$$

Next, suppose $\mathbb{T}$ is not symmetrical. Then we consider $\widehat{\mathbb{T}} = \mathbb{T} \cup (-\mathbb{T})$ with $-\mathbb{T} = \{-t \colon t \in \mathbb{T}\}$. Clearly, $\widehat{\mathbb{T}}$ is symmetrical about zero. Let

$$
G(t) = \begin{cases} F(t) & \text{for } t \in \mathbb{T} \\ F(0) & \text{for } t \in \widehat{\mathbb{T}} \setminus \mathbb{T}. \end{cases}
$$

Hence, G is regulated and almost periodic on $\widehat{\mathbb{T}}$. Thus, in view of the above discussion, we obtain that G is bounded, subsequently, F is bounded. This completes the proof. $\qquad\square$

In what follows, we shall prove that the Δ_g-derivative and the Δ_g pre-antiderivative of an almost periodic set-valued function are almost periodic. We need the following mean value theorem for single-valued real functions defined on $\mathbb{T}$.

Theorem 3.21. *Let* $f \colon \mathbb{T} \to \mathbb{R}$ *be pre-differentiable with* $\mathbf{D}$. *If* $U \subset \mathbb{T}$ *is a compact interval with endpoints* $r, s \in \mathbb{T}$, *then*

$$
\left\{ \sup_{t \in U \cap \mathbf{D}} f^{\Delta}(t) \right\} (s - r) \geq f(s) - f(r) \geq \left\{ \inf_{t \in U \cap \mathbf{D}} f^{\Delta}(t) \right\} (s - r)
$$

for all $s, r \in \mathbb{T}$ *with* $r \leq s$.

Proof. Define

$$
F(t) = \left\{ \inf_{t \in U \cap \mathbf{D}} f^{\Delta}(t) \right\} (t - r) \quad \text{for } t \in \mathbb{T}.
$$

Then,

$$
F^{\Delta}(t) = \left\{ \inf_{t \in U \cap \mathbf{D}} f^{\Delta}(t) \right\} \leq f^{\Delta}(t) \text{ for } t \in [r, s] \cap \mathbf{D}.
$$

From Theorem 1.2, it follows that

$$
f(t) - f(r) \geq F(t) - F(r) \quad \text{for } t \in [r, s]
$$

so that

$$
f(s) - f(r) \geq F(s) - F(r) = F(s) = \left\{ \inf_{t \in U \cap \mathbf{D}} f^{\Delta}(t) \right\} (s - r).
$$

Similarly, we can prove the first inequality. This completes the proof. $\qquad\square$

The following lemma is an immediate extension of corresponding result for the single-valued functions on $\mathbb{R}$.

Theorem 3.22. *Let $f \in AP(\mathbb{T}, \mathbb{R})$ be a single-valued delta differentiable function such that f^{Δ} is rd-continuous. Then, $f^{\Delta} \in AP(\mathbb{T}, \mathbb{R})$ if and only if it is uniformly rd-continuous on $\mathbb{T}$.*

Proof. In view of Proposition 3.1(a2), if $f^{\Delta} \in AP(\mathbb{T}, \mathbb{R})$, then the rd-continuity of f^{Δ} implies its uniform rd-continuity. It is sufficient to verify that the uniform rd-continuity of f^{Δ} infers its almost periodicity. Let $\{h_m\}$ be a sequence of positive reals such that $h_m \to 0$ as m tends to infinity. For each $m \in \mathbb{N}$, consider the mapping i_m defined on $\mathbb{T}$ as

$$
i_m(t) \begin{cases} \in (0, h_m] \text{ with } t + i_m(t) \in \mathbb{T} & \text{if } t \text{ is right-dense} \\ = \mu(t) & \text{if } t \text{ is right-scattered.} \end{cases}
$$

Denote

$$
f_m(t) = i_m^{-1}(t)[f(t + i_m(t)) - f(t)] \quad \text{for } t \in \mathbb{T} \quad \text{and} \quad m \in \mathbb{N}.
$$

It is easy to see that, for $t \in \mathbb{T}$, $f_m(t) \to f^{\Delta}(t)$ as $m \to \infty$. From Theorem 3.21, it follows that

$$
\left\{ \sup_{s \in [t, t + i_m(t)]} f^{\Delta}(s) \right\} - f^{\Delta}(t)
$$

$$
\geq f_m(t) - f^{\Delta}(t) \tag{3.55}
$$

$$
\geq \left\{ \inf_{s \in [t, t + i_m(t)]} f^{\Delta}(s) \right\} - f^{\Delta}(t) \quad \text{for } t \in \mathbb{T}.
$$

Since f^{Δ} is uniformly rd-continuous on $\mathbb{T}$, for given $\varepsilon > 0$, there exists $\delta = \delta(\varepsilon) > 0$ such that for any $t, s \in \mathbb{T}$ with t right-dense and $|t - s| < \delta$, we have

$$
|f^{\Delta}(t) - f^{\Delta}(s)| < \varepsilon. \tag{3.56}
$$

We now take $n_0 \in \mathbb{N}$ large enough such that $h_m < \delta$ for each $m \geq n_0$. For any $t \in \mathbb{T}$, we have $|s - t| < \delta$ as long as $s \in [t, t + h_m]$ for $m \in \mathbb{N}$. If t is right-dense, in virtue of (3.55) and (3.56), we have

$$
\varepsilon > f_m(t) - f^{\Delta}(t) > -\varepsilon \quad \text{for each } t \in \mathbb{T}. \tag{3.57}
$$

If t is right-scattered element of $\mathbb{T}$, then $f_m(t) = f^{\Delta}(t)$. Thus, (3.57) is satisfied for all $t \in \mathbb{T}$, that is, f_m uniformly converges to f^{Δ} on $\mathbb{T}$ as m

goes to infinity. Now, Proposition 3.1(a5) guarantees $f^\Delta \in AP(\mathbb{T}, \mathbb{R})$. This completes the proof. □

Now, in order to extends the result of Theorem 3.22 into the case of set-valued functions, we recall the concept of the support functions introduced by Lakshmikantham *et al.* (2006).

Definition 3.13. Let $\mathcal{A} \subset \mathbb{R}^n$. The support function of $\mathcal{A}$ is defined, for any $p \in \mathbb{R}^n$, by

$$s(p, \mathcal{A}) = \sup\{\langle p, a \rangle : a \in \mathcal{A}\}, \tag{3.58}$$

which may take the value $+\infty$ when $\mathcal{A}$ is unbounded. However, if $\mathcal{A} \in K_c^n$, then $s(\cdot, \mathcal{A})$ is well-defined and $s(p, \mathcal{A}) = s(p, \mathcal{B})$ for all $p \in \mathbb{R}^n$, if and only if $\mathcal{A} = \mathcal{B}$.

The support function s has the following properties:

(SP1) For $\mathcal{A}, \mathcal{B} \subset \mathbb{R}^n$ and $p \in \mathbb{R}^n$,

$$s(p, \mathcal{A} + \mathcal{B}) = s(p, \mathcal{A}) + s(p, \mathcal{B})$$

and

$$s(p, t\mathcal{A}) = ts(p, \mathcal{A}) \quad \text{when } t \geq 0.$$

(SP2) Let $S^{n-1} := \{p \in \mathbb{R}^n : \|p\| = 1\}$. For $\mathcal{A}, \mathcal{B} \in K_c^n$, we have

$$D[\mathcal{A}, \mathcal{B}] = \sup\{|s(p, \mathcal{A}) - s(p, \mathcal{B})| : p \in S^{n-1}\}.$$

Thus, $\mathcal{A} = \mathcal{B}$ if and only if $s(p, \mathcal{A}) = s(p, \mathcal{B})$ for each $p \in S^{n-1}$.

(SP3) Let $F : \mathbb{T} \to K_c^n$. Then, for $t, t' \in \mathbb{T}$ and $p, p' \in \bar{S}^n = \{p \in \mathbb{R}^n : \|p\| \leq 1\}$, we have

$$|s(p, F(t)) - s(p', F(t'))| \leq \|F(t)\|\|p - p'\| + D[F(t), F(t')].$$

Keeping in mind the properties of the Hausdorff metric and (SP2), we obtain

$$D[\mathcal{A}, \mathcal{B}] = D[\mathcal{A} -_g \mathcal{B}, \{0\}]$$

$$= \sup\{|s(p, \mathcal{A} -_g \mathcal{B}) - s(p, \{0\})| : p \in S^{n-1}\}$$

$$= \sup\{|s(p, \mathcal{A} -_g \mathcal{B})| : p \in S^{n-1}\}$$

provided the g-difference of $\mathcal{A}$ and $\mathcal{B}$ exists, It is easy to see that

$$|s(p, \mathcal{A}) - s(q, \mathcal{A})| \le \|\mathcal{A}\| \|p - q\|$$

for all $p, q \in \mathbb{R}^n$. This implies that $s(p, \mathcal{A})$ is continuous with respect to p for given $\mathcal{A} \in K_c^n$. Since S^{n-1} is compact, there exists $p_0 \in S^{n-1}$ such that, for $\mathcal{A}, \mathcal{B} \in K_c^n$,

$$
\begin{aligned}
D[\mathcal{A}, \mathcal{B}] &= |s(p_0, \mathcal{A} -_g \mathcal{B})| \\
&= |s(p_0, \mathcal{A}) - s(p_0, \mathcal{B})|.
\end{aligned}
\tag{3.59}
$$

Lemma 3.3. *Let $F : \mathbb{T} \to K_c^n$ and $p \in \mathbb{R}^n$ and $j_{F,p}(t) = s(p, F(t))$, $t \in \mathbb{T}$. Then,*

(i) *the function $j_{F,p}$ is continuous (resp., rd-continuous, uniformly continuous) if and only if F is continuous (resp., rd-continuous, uniformly continuous) on $\mathbb{T}$;*

(ii) *the almost periodicity of $j_{F,p}$, for any $p \in S^{n-1}$, is equivalent to the almost periodicity of F;*

(iii) *if F is Δ_g-differentiable on $\mathbb{T}$, then $j_{F,p}$ is Δ-differentiable on $\mathbb{T}$ for any $p \in S^{n-1}$. Moreover,*

$$j_{F,p}^{\Delta}(t) = s(p, \Delta_g F(t)) \quad \text{for } t \in \mathbb{T}.$$

Proof. The proof of (i) is trivial by (SP3) and (3.59). To prove (ii), assume that F is almost periodic. For given $\varepsilon > 0$, let $l = l(\varepsilon)$ be the inclusion length of $\mathscr{E}(\varepsilon, F)$ (which is defined in (3.47)) and $\omega \in C^*$ is an ε-almost period of F. By means of (SP3), for $t \in \mathbb{T}$ and $p \in S^{n-1}$, we have

$$|j_{F,p}(t + \omega) - j_{F,p}(t)| \le D[F(t + \omega), F(t)]$$

$$< \varepsilon.$$

This shows that $j_{F,p}$ is almost periodic. Conversely, assume that $j_{F,p}$ is almost periodic for any given $p \in \mathbb{T}$. For any $\varepsilon > 0$, let $l = l(\varepsilon)$ be the inclusion length of $\mathscr{E}(\varepsilon, j_{F,p_0})$ and $\omega \in C^*$ is an ε-almost period of j_{F,p_0}. By means of (3.59), for any $t \in \mathbb{T}$, we have

$$D[F(t + \omega), F(t)] = |s_{F,p_0}(t + \omega) - s_{F,p_0}(t)|$$

$$< \varepsilon.$$

This shows that F is almost periodic. Hence, (ii) holds.

Finally, we prove (iii). Given $t \in \mathbb{T}$ and $p \in S^{n-1}$ and for any $\varepsilon > 0$, if F is Δ_g-differentiable, then there exists a neighborhood $U_\mathbb{T}$ of t such that

$$|j_{F,p}(t+h) - j_{F,p}(\sigma(t)) - s(p, \Delta_g F(t))(h - \mu(t))|$$
$$= |j_{F,p}(t+h) - [s(p, F(\sigma(t))) + s(p, (h - \mu(t))\Delta_g F(t))]|$$
$$= |s(p, F(t+h)) - s(p, F(\sigma(t))) - s(p, (h - \mu(t))\Delta_g F(t))|$$
$$\leq D[F(t+h), F(\sigma(t)) - s(p, (h - \mu(t))\Delta_g F(t)]$$
$$= D[F(t+h) -_g F(\sigma(t)), (h - \mu(t))\Delta_g F(t)]$$
$$\leq \varepsilon|h - \mu(t)|.$$

That is,

$$|j_{F,p}(t+h) - j_{F,p}(\sigma(t)) - s(p, \Delta_g F(t))(h - \mu(t))| \leq \varepsilon|h - \mu(t)|$$

for all $t + h \in U_\mathbb{T}$ with $|h| < \delta$. This completes the proof. $\square$

We are now in a position to verify the almost periodicity of the Δ_g-derivative and the integrals of almost periodic set-valued functions and corresponding results on $\mathbb{T}$.

Theorem 3.23. *Let* $F \colon \mathbb{T} \to K_c^n$ *be such that* $F \in AP(\mathbb{T}, K_c^n) \cap C_{rd}^1(\mathbb{T}, K_c^n)$. *Then,* $\Delta_g F \in AP(\mathbb{T}, K_c^n)$ *if and only if it is uniformly rd-continuous on* $\mathbb{T}$.

Proof. Let $j_{F,p},(t) = s(p, F(t))$ be the support function of $F(t)$, $t \in \mathbb{T}$. Then, Lemma 3.3(ii) implies $j_{F,p} \in AP(\mathbb{T}, \mathbb{R})$ for all $p \in S^{\tau-1}$. Under our hypothesis, from Lemma 3.3(i) and (iii), it follows that $j_{F,p}$ is continuous, delta differentiable and its delta derivative $j_{F,p}^\Delta$ is the support function of $\Delta_g F$. Moreover, $j_{F,p}^\Delta$ is uniformly continuous on $\mathbb{T}$. In virtue of Theorem 3.22, we get

$$j_{\Delta_g F, p} = j_{F,p}^\Delta \in AP(\mathbb{T}, \mathbb{R}) \quad \text{for all } p \in S^{n-1}.$$

Lemma 3.3(ii) guarantees that $\Delta_g F \in AP(\mathbb{T}, K_c^n)$. This completes the proof. $\square$

Theorem 3.24. *Let* $F \in AP(\mathbb{T}, K_c^n)$. *Then the function* $\mathfrak{F} \colon \mathbb{T} \to K_c^n$ *defined by*

$$\mathfrak{F}(t) = \int_a^t F(s)\Delta_g s$$

with $a \in \mathbb{T}$, *is almost periodic if and only if* $\mathfrak{F}$ *is bounded.*

Proof. Let $j_{\mathfrak{F},p}(t) = s(p, \mathfrak{F}(t))$ be the support function of $\mathfrak{F}(t)$, $t \in \mathbb{T}$. By means of Theorem 2.1(i), we see that $\mathfrak{F}$ is Δ_g-differentiable and

$$\Delta_g \mathfrak{F}(t) = F(t) \quad \text{for each } t \in \mathbb{T}.$$

In the view of Lemma 3.3(iii), we obtain that $j_{\mathfrak{F},p}$ is delta differentiable and

$$j_{\mathfrak{F},p}^{\Delta}(t) = s(p, \Delta_g \mathfrak{F}(t)) = s(p, F(t))$$

for all $t \in \mathbb{T}$ and $p \in S^{n-1}$. In view of the definition of Cauchy integral for single-valued function, we have

$$\int_a^t s(p, F(s)) \Delta s = \int_a^t j_{\mathfrak{F},p}^{\Delta}(s) \Delta s$$
$$= j_{\mathfrak{F},p}(t) - j_{\mathfrak{F},p}(a)$$
$$= j_{\mathfrak{F},p}(t).$$

Since $F \in AP(\mathbb{T}, K_c^n)$, from Lemma 3.3(ii), it follows that $s(\cdot, F(t))$ is almost periodic. According to the well known property of single-valued functions, the delta antiderivative $j_{\mathfrak{F},p}(\cdot)$ of $s(p, F(\cdot))$ is almost periodic if and only if it is bounded. Hence, keeping in mind Lemma 3.3(ii), it is sufficient to verify that, for any $p \in S^{n-1}$, the boundedness of $j_{\mathfrak{F},p}$ and $\mathfrak{F}$ are equivalent. Indeed, in view of (SP2) and (3.59), this equivalence is true. This completes the proof. $\qquad\square$

3.5.2 *Almost periodicity of fuzzy functions*

In this section, we discuss the almost periodicity of fuzzy number-valued functions. For this, we need some terse memories for the modification of fuzzy theory on the base of Bede and Gal (2004) and Khastan *et al.* (2011). A function $u\colon \mathbb{T} \to [0,1]$ is called a fuzzy number on time scale provided it satisfies the following properties:

(i) u is normal, i.e., there exists $s_0 \in \mathbb{T}$ such that $u(s_0) = 1$;
(ii) u is a convex fuzzy set on $\mathbb{T}$, that is, for all $t \in [0,1]$, $a, b \in \mathbb{T}$, satisfying $ta + (1-t)b \in \mathbb{T}$, we have

$$u(ta + (1-t)b) \geq \min\{u(a), u(b)\};$$

(iii) u is upper semicontinuous on $\mathbb{T}$; and

(iv) The set $[u]^0 = \overline{\{s \in \mathbb{T}: u(s) > 0\}}$ is compact, where $\overline{A}$ denotes the closure of A in $\mathbb{R}$.

Let $\mathbb{T}_f$ denote the space of fuzzy numbers. For $0 < r \leq 1$, the set

$$[u]^r = \{s \in \mathbb{T}: u(s) \geq r\}$$

is called the r-level set of u. Obviously, $[u]^r$ is a compact interval of $\mathbb{T}$ if $u \in \mathbb{T}_f$ for all $r \in [0, 1]$. The notation $[u]^r = [u_r^-, u_r^+]$ denotes explicitly the r-level set of u. We refer to u^- and u^+ as the lower and upper branches of u, respectively. For $u, v \in \mathbb{T}_f$ and $\lambda \in \mathbb{R}$, the addition $u + v$ and the scalar multiplication λu have r-level sets

$$[u + v]^r = [u]^r + [v]^r$$

and

$$[\lambda u]^r = \lambda [u]^r$$

for any $r \in [0, 1]$, respectively, where $[u]^r + [v]^r$ is defined as the same as the usual addition of two intervals (subsets) of $\mathbb{T}$ and $\lambda [u]^r$ means the usual product between a scalar and a subset of $\mathbb{T}$. Note that $u + v \in \mathbb{T}_f$ and $\lambda u \in \mathbb{T}_f$.

The g-difference for fuzzy numbers can be defined as follows. For $u, v \in \mathbb{T}_f$, the g-difference is the fuzzy number w, if it exists, such that

$$u -_g v = w \Leftrightarrow \begin{cases} \text{(a)} & u = v + w \quad \text{or} \\ \text{(b)} & v = u + (-1)w. \end{cases} \tag{3.60}$$

If $u -_g v$ exists, then its r-level sets are given by

$$[u -_g v]^r = [\min\{u_r^- - v_r^-, u_r^+ - v_r^+\}, \max\{u_r^- - v_r^-, u_r^+ - v_r^+\}]$$

and $u -_H v = u -_g v$ provided $u -_H v$ exists. If (a) and (b) of (3.60) are satisfied simultaneously, then w is a crisp number. Also, $u -_g u = u -_H u = \theta$.

For any $u, v \in \mathbb{T}_f$, the metric structure is given by the Hausdorff distance

$$D_f[u, v] = \sup_{r \in [0,1]} \max\{|u_r^- - v_r^-|, |u_r^+ - v_r^+|\}.$$

Thus, $(\mathbb{T}_f, D_f)$ is a complete metric space. Under the consideration of this distance D_f and g-difference, the definitions of continuity and g-derivative of a fuzzy number-valued function are similar to one in the previous paragraph. By using this Hausdorff distance D_f, Bede and Gal in Bede and Gal (2004) have developed the theory of almost periodic fuzzy functions on $\mathbb{R}$. In this sequel, we will extend this concept to fuzzy functions from $\mathbb{T}$ into $\mathbb{T}_f$. First, we have the following.

Definition 3.14. Let $\mathbb{T}$ be an almost periodic time scale with respect to $\mathcal{D}$. A fuzzy number-valued function $f \in \mathcal{C}_{\mathrm{rd}}(\mathbb{T}, \mathbb{T}_f)$ is called almost periodic provided for any given $\varepsilon > 0$, the set

$$\mathscr{E}(\varepsilon, f) = \{\omega \in C^* \colon D_f[f(t+\omega), f(t)] < \varepsilon \quad \text{for all } t \in \mathbb{T}\}$$

is relatively dense in $\mathbb{T}$. That is, there exists an $l = l(\varepsilon) > 0$ such that each interval of length l contains at least one $\omega = \omega(\varepsilon) \in \mathscr{E}(\varepsilon, F)$, that is, $\omega \in C^*$ and

$$D_f[f(t+\omega), f(t)] < \varepsilon \quad \text{for all } t \in \mathbb{T}.$$

The set $\mathscr{E}(\varepsilon, f)$ is called an ε-almost periodic set of f, ω is called an ε-almost period of f, and l is called the inclusion length of $\mathscr{E}(\varepsilon, f)$. The collection of all almost periodic functions from $\mathbb{T}$ into $\mathbb{T}_F$ is denoted by $AP(\mathbb{T}, \mathbb{T}_f)$.

Consider the r-level set of f given by

$$[f(t)]^r = [f_r^-(t), f_r^+(t)] \cap \mathbb{T}$$

for each $t \in \mathbb{T}$. From Definition 3.14, we have that $f \in AP(\mathbb{T}, \mathbb{T}_f)$ if and only if

$$\{\omega \in C^* \colon |g_r(t+\omega) - g_r(t)| < \varepsilon \text{ for all } t \in \mathbb{T} \text{ and } r \in [0,1]\}$$

is relatively dense in $\mathbb{T}$ with $g_r = f_r^-$ or $g_r = f_r^+$. That is, the lower and upper branches of f, viz. f^- and f^+, are almost periodic in $t \in \mathbb{T}$ uniformly for $r \in [0, 1]$.

Note that $[u]^r \in K_c^1$ provided $u \in \mathbb{T}_f$ for $r \in [0, 1]$. By similar arguments, we can obtain the corresponding results as in Theorems 3.20, 3.23 and 3.24, also, in Proposition 3.1 for fuzzy functions on $\mathbb{T}$, which improve and unify corresponding problems for fuzzy functions on the continuous system ($\mathbb{T} = \mathbb{R}$ discussed in Bede and Gal (2004)) and on the discrete system ($\mathbb{T} = \mathbb{Z}$).

3.5.3 *Existence of almost periodic solutions*

In this section, we use the exponential dichotomy of SDEs given in Definition 3.10, to discuss an almost periodic solutions of SDEs. Let $0 \in \mathbb{T}$. For the sake of convenience, in this section, we restrict our discussions to $\mathbb{T}^+$ and we shall show that is possible without loss of generality.

Lemma 3.4. *Suppose that* (3.32) *admits an exponential dichotomy on* $[t_0, +\infty)_\mathbb{T}$ *with* $t_0 \in \mathbb{T}^+$, *so does it on* $\mathbb{T}^+$ *with the same constant a and projection* P.

Proof. Let $s, t \in \mathbb{T}^+$. Take a constant $\lambda \geq 1$ such that
$$\left| e_A(u, 0) e_{\ominus A^*}^*(v, 0) \right| \leq \lambda$$
for any u, v satisfying $0 \leq u, v \leq \sigma(t_0)$. For $t \geq t_0 \geq \sigma(s) \geq 0$, we have
$$\left| e_A(t, 0) P e_{\ominus A^*}^*(s, 0) \right| = \left| e_A(t, 0) P e_{\ominus A^*}^*(t_0, 0) e_A(t_0, 0) e_{\ominus A^*}^*(s, 0) \right|$$
$$\leq \lambda k e_{\ominus a}(t, \sigma(t_0))$$
$$= \lambda k e_{\ominus a}(t, \sigma(s)) e_{\ominus a}(\sigma(s), \sigma(t_0))$$
$$\leq \lambda^2 k e_a(\sigma(t_0), 0) e_{\ominus a}(t, \sigma(s)).$$
Also, for $t_0 \geq t \geq \sigma(s) \geq 0$, we have
$$\left| e_A(t, 0) P e_{\ominus A^*}^*(s, 0) \right|$$
$$= \left| e_A(t, 0) e_{\ominus A^*}^*(t_0, 0) e_A(t_0, 0) P e_{\ominus A^*}^*(t_0, 0) e_A(t_0, 0) e_{\ominus A^*}^*(s, 0) \right|$$
$$\leq \lambda^2 k e_{\ominus a}(t_0, \sigma(t_0))$$
$$= \lambda^2 k e_{\ominus a}(t_0, t) e_{\ominus a}(t, \sigma(s)) e_{\ominus a}(\sigma(s), \sigma(t_0))$$
$$\leq \lambda^2 k e_a(t_0, t_0) e_{\ominus a}(t, \sigma(s)) e_a(\sigma(t_0), 0)$$
$$= \lambda^2 k e_a(\sigma(t_0), 0) e_{\ominus a}(t, \sigma(s)).$$
Hence, for $0 \leq s \leq t < +\infty$, we have
$$\left| e_A(t, 0) P e_{\ominus A^*}^*(s, 0) \right| \leq \hat{k} e_{\ominus a}(t, \sigma(s)),$$
where $\hat{k} = \lambda^2 k e_a(\sigma(t_0), 0)$. In the case of $0 \leq t \leq s < \infty$, by the similar argument, we obtain
$$\left| e_A(t, 0)(\mathcal{I} - P) e_{\ominus A^*}^*(s, 0) \right| \leq \hat{k} e_{\ominus a}(\sigma(s), t).$$
This completes the proof. $\qquad\square$

Consider the set dynamic equation

$$\Delta_g U(t) = A(t)U(t) + F(t) \quad \text{for } t \in \mathbb{T}, \tag{3.61}$$

where $A \in \mathcal{R}_n$ and $F \in \mathcal{C}_{\mathrm{rd}}(\mathbb{T}, K_c^n)$. Note that (3.61) is the first equation in (3.28). Let $B \in H(A)$ and $G \in H(F)$ with A and F given as in (3.61). We call the equations

$$\Delta_g V = B(t)V \tag{3.62}$$

and

$$\Delta_g V = B(t)V + G(t) \tag{3.63}$$

a homogeneous equation in the hull and a nonhomogeneous equation in the hull of (3.61), respectively. Here, $V \in \mathcal{C}_{\mathrm{rd}}(\mathbb{T}, K_c^n)$. The following results are due to the idea of Coppel discussed in Coppel (1978).

Theorem 3.25. *Let A be rd-continuous on $\mathbb{T}$ and (3.32) admits an exponential dichotomy on $\mathbb{T}^+$ with constants k, a and the projection P. Suppose there exists a sequence $\alpha = \{\alpha_m\} \subset C^*$ satisfying $\alpha_m \to \infty$ as $m \to \infty$ such that*

$$T_\alpha A(t) = B(t)$$

uniformly on each compact interval of $\mathbb{T}$. Then the homogeneous equation in the hull (3.62) admits an exponential dichotomy on $\mathbb{T}$ with the same constants k, a and the projection

$$Q = \lim_{m \to \infty} e_{\ominus A^*}^*(\alpha_m, 0)P e_A(\alpha_m, 0).$$

Proof. Note that the fundamental matrix $e_A^{(m)}(t, 0)$ of the translation

$$\Delta_g U = A(t + \alpha_m)U$$

satisfies

$$\begin{aligned} e_A^{(m)}(t, 0) &= e_A(t + \alpha_m, \alpha_m) \\ &= e_A(t + \alpha_m, 0)e_A(0, \alpha_m) \\ &= e_A(t + \alpha_m, 0)e_{\ominus A^*}^*(\alpha_m, 0). \end{aligned}$$

In view of the exponential dichotomy of (3.32), for $t, s \in \mathbb{T}$, we have

$$\left| e_A(t + \alpha_m, 0)(\mathcal{I} - P_m)e^*_{\ominus A^*}(s + \alpha_m, 0) \right| \leq k e_{\ominus a}(\sigma(s), t)$$

when $s \geq t \geq -\alpha_m$, and

$$\left| e_A(t + \alpha_m, 0)P_m e^*_{\ominus A^*}(s + \alpha_m, 0) \right| \leq k e_{\ominus a}(t, \sigma(s))$$

when $t \geq \sigma(s) \geq -\alpha_m$, where

$$P_m = e^*_{\ominus A^*}(\alpha_m, 0)Pe_A(\alpha_m, 0).$$

It is clear that, for each $m \in \mathbb{N}$, P_m is also a projection. Since $|P_m| \leq k$ for $m \in \mathbb{N}$, the sequence $\{P_m\}$ has a convergent subsequence. Without loss of generality, we assume that $\{P_m\}$ is also convergent and

$$\lim_{m \to \infty} P_m = Q.$$

Hence, Q is a projection too.

Let $X_m(t) = X(t + \alpha_m)X^{-1}(\alpha_m)$ be the fundamental solution matrix of the linear equation

$$x^\Delta(t) = A(t + \alpha_m)x$$

with $x \in \mathbb{R}^n$. Then, in virtue of Lemma A.3, we have

$$X_m(t) \to Y(t) \quad \text{as } m \to \infty, \quad \text{for each } t \in \mathbb{T},$$

where $Y(t)$ is a fundamental solution matrix of the homogeneous linear equation in the hull

$$y^\Delta(t) = B(t)y$$

with $Y(0) = \mathcal{I}$. On the other hand, we have

$$X_m(t) = e_A(t + \alpha_m, \alpha_m)X(\alpha_m) \quad \text{and} \quad Y(t) = e_B(t, 0)Y(0).$$

This implies that

$$e_A(t + \alpha_m, 0) \to e_B(t, 0) \quad \text{as } m \to \infty,$$

and

$$\left| e_B(t, 0)Qe^*_{\ominus B^*}(s, 0) \right| \leq k e_{\ominus a}(t, \sigma(s)) \quad \text{for } t, s \in \mathbb{T} \text{ with } t \geq \sigma(s),$$

and

$$\left| e_B(t, 0)(\mathcal{I} - Q)e^*_{\ominus B^*}(s, 0) \right| \leq k e_{\ominus a}(\sigma(s), t) \quad \text{for } t, s \in \mathbb{T} \text{ with } s \geq t.$$

This completes the proof. $\qquad\square$

By means of Proposition 3.25, we immediately infer the following.

Proposition 3.2. *Let A be an almost periodic matrix-valued function. Then the exponential dichotomy of (3.32) on $\mathbb{T}^+$ and that on $\mathbb{T}$ is equivalent.*

Recall that (3.61) is said to satisfy the Favard condition provided each nontrivial bounded solution V of (3.62), a homogeneous equation in the hull of (3.61), satisfies

$$\inf_{t \in \mathbb{T}} \|V(t)\| > 0.$$

Let $\mathscr{B}$ denotes the set consisting of bounded solutions to the nonlinear set dynamic equation

$$\Delta_g U = F(t, U), \tag{3.64}$$

where $F \in \mathcal{C}(\mathbb{T} \times K_c^n, K_c^n)$. If $\mathscr{B} \neq \emptyset$, then

$$\underline{\lambda} = \inf_{U \in \mathscr{B}} \|U\|$$

exists, where

$$\|U\| = \sup_{t \in \mathbb{T}} \|U(t)\|.$$

This $\underline{\lambda}$ is called the least-value of solutions to (3.64). In this sense, V is called a minimum norm solution to (3.64) provided V belongs to $\mathscr{B}$ such that $\|V\| = \underline{\lambda}$.

Theorem 3.26. *Let the sequence $\{U_m\}$ be Δ_g-differentiable on $\mathbb{T}$ for $m \in \mathbb{N}$. Suppose $\{U_m\}$ and $\{\Delta_g U_m\}$ are uniformly bounded on $\mathbb{T}$, i.e., there exist constants $M, M' > 0$ such that*

$$\|U_m(t)\| \leq M \quad and \quad \|\Delta_g U_m(t)\| \leq M' \quad for\ all\ m\ and\ t \in \mathbb{T}.$$

Then, $\{U_m\}$ is equicontinuous on $\mathbb{T}$. Moreover, there exists a subsequence of $\{U_m\}$ that is uniformly convergent on any compact subset of $\mathbb{T}$.

Proof. Let $j_{m,p}(t) = s(p, U_m(t))$, where $s(p, U_m(t))$ is the support function of $U_m(t)$ as defined in Definition 3.13. From Lemma 3.3(iii), it follows that $j_{m,p}$ is Δ-differentiable and

$$j_{m,p}^{\Delta}(t) = s(p, \Delta_g U_m(t)).$$

According to Corollary 1.1, for $h > 0$, we have

$$|j_{m,p}(t+h) - j_{m,p}(t-h)| \leq 2h \left\{ \sup_{s \in [t-h, t+h]_{\mathbb{T}}} |j_{m,p}^{\Delta}(s)| \right\}.$$

Since $\|\Delta_g U_m(t)\| = \sup \left\{ |j_{m,p}^{\Delta}(t)| : p \in S^{\tau-1} \right\}$, from our assumptions, we obtain

$$|j_{m,p}^{\Delta}(s)| \leq M'$$

for $m \in \mathbb{N}$, $p \in S^{n-1}$ and $s \in [t-h, t+h]_{\mathbb{T}}$. Substituting this in the above inequality yields

$$D[U_m(t+h), U_m(t-h)] = \sup \left\{ |j_{m,p}(t+h) - j_{m,p}(t-h)| \right\}$$
$$\leq 2M'h$$

for all $m \in \mathbb{N}$. This gives the equicontinuity of $\{U_m\}$.

If $\{U_m\}$ is uniformly bounded on $\mathbb{T}$, by the Arzelà–Ascoli theorem for set-valued mappings, given in Theorem A.5, there exists a subsequence $\{U_{m_l^{(1)}}\}$ which is uniformly convergent on the interval $I_1 = [-1,1]_{\mathbb{T}}$. We can inductively find the subsequence $\{m^{(i+1)}\}$ of $\{m^{(i)}\}$ such that $\{U_{m_l^{(i+1)}}\}$ is uniformly convergent on $[-(i+1), i+1]_{\mathbb{T}}$ for $i \in \mathbb{N}$. We now take the subsequence $\{m_i^{(i)}\}$ of diagonal elements. Then, $\{U_{m_i^{(i)}}\} \subset \{U_m\}$ is uniformly convergent on each interval $[-i, i]_{\mathbb{T}}$ for $i \in \mathbb{N}$. Since every compact subset of $\mathbb{T}$ must be included in some of the interval $[-m, m]_{\mathbb{T}}$, the desired result achieves. This completes the proof. $\square$

The following theorems will be useful and whose proofs are close analogous to the corresponding classical case.

Theorem 3.27. *Let $\mathscr{S} \subset K_c^n$ be a closed ball with $\theta \in \mathscr{S}$ and $\{U(t) : t \in \mathbb{T}\} \subset \mathscr{S}$ for $U \in \mathscr{B}$. If $F \in C(\mathbb{T} \times \mathscr{S}, K_c^n)$ is bounded and $\mathscr{B} \neq \emptyset$, then (3.64) has a minimum norm solution.*

Proof. The proof is analogous to that of Lemma A.4. $\square$

Theorem 3.28. *If $F \in AP_u(\mathbb{T} \times K_c^n, K_c^n)$ and (3.64) has a bounded solution U_0 on $[t_0, \infty)_\mathbb{T}$, then there exists $V \in \mathscr{B}$ such that*

$$\{V(t) \colon t \in \mathbb{T}\} \subset \overline{\{U_0(t) \colon t \geq t_0,\ t \in \mathbb{T}\}}.$$

Proof. The proof is analogous to that of Lemma A.5. $\qquad \square$

Theorem 3.29. *Let $F \in AP_u(\mathbb{T} \times K_c^n, K_c^n)$ be continuous. Suppose, for each $G \in H(F)$, the equation in the hull of (3.64),*

$$\Delta_g V = G(t, V)$$

has a unique minimum norm solution V. Then V is almost periodic on $\mathbb{T}$.

Proof. The proof is analogous to that of Lemma A.6. $\qquad \square$

The following existence result of almost periodic solutions for linear nonhomogeneous set dynamic equation generalizes the well-known Favard theorem.

Theorem 3.30. *Let A be an $n \times n$-almost periodic matrix-valued function and $F \in AP(\mathbb{T}, K_c^n)$ be such that (3.61) has a bounded solution on $[t_0, \infty)_\mathbb{T}$. If (3.61) satisfies Favard condition, then (3.61) has an almost periodic solution.*

Proof. From Theorems 3.27 and 3.28, it follows that (3.61) has bounded and minimum norm solutions on $\mathbb{T}$. Moreover, for any $B \in H(A)$ and $G \in H(F)$, the nonhomogeneous equation in the hull, (3.63) has also a minimum norm solution. We shall verify that, for any given $B \in H(A)$ and $G \in H(F)$, (3.63) has a unique minimum norm solution. In fact, let $B \in H(A)$ and $G \in H(F)$ be fixed and U_1, U_2 both be the minimum norm solutions of (3.63) with the least-value λ. From (3.63), it follows that

$$\Delta_g U_1 -_g B(t) U_1 = \Delta_g U_2 -_g B(t) U_2 = G(t) \quad \text{for } t \in \mathbb{T}.$$

By Proposition 2.4 (d1), this yields

$$\Delta_g(U_1 -_g U_2) = B(t)(U_1 -_g U_2),$$

which further shows that $\frac{1}{2}(U_1 -_g U_2)$ is a solution of (3.62). If $U_1 \neq U_2$, then, in virtue of the Favard condition, there exists $\rho > 0$

such that

$$\inf_{t \in \mathbb{T}} \|U_1(t) -_g U_2(t)\| \geq 2\rho > 0.$$

For any given $t \in \mathbb{T}$, denote

$$V(t) = \{u_1(t) - u_2(t) \colon u_1(t) \in U_1(t),\ u_2(t) \in U_2(t)\},$$

and

$$U(t) = U_1(t) -_g U_2(t).$$

We assert that $U(t) \subset V(t)$. Indeed, let us fix $u(t) \in U(t)$. If g-difference exists in the sense (a) of (2.1), then

$$U_1(t) = U_2(t) + U(t).$$

For any $u_2(t) \in U_2(t)$, we have $u_1(t) \in U_1(t)$, with $u_1(t) = u(t) + u_2(t)$. That is,

$$u(t) \in V(t) \quad \text{with } u(t) = u_1(t) - u_2(t).$$

If g-difference exists in the sense (b) of (2.1), then

$$U_2(t) = U_1(t) + (-1)U(t).$$

Take $u_1(t) \in U_1(t)$. Since $-u(t) \in (-1)U(t)$, we get

$$u_2(t) \in U_2(t) \quad \text{with } u_2(t) = u_1(t) - u(t).$$

This yields that

$$u(t) \in V(t) \quad \text{with } u(t) = u_1(t) - u_2(t).$$

Consequently, $U(t) \subset V(t)$ which deduces our assertion.

On the other hand, for any $u_i(t) \in U_i(t)$ with $i = 1, 2$ and $t \in \mathbb{T}$, by means of the parallelogram law, we have

$$\left|\frac{1}{2}(u_1(t) + u_2(t))\right|^2 + \left|\frac{1}{2}(u_1(t) - u_2(t))\right|^2 = \frac{1}{2}\left(|u_1(t)|^2 + |u_2(t)|^2\right)$$

$$\leq \lambda^2.$$

This implies that

$$\left\| \frac{1}{2}\left(U_1(t) + U_2(t)\right) \right\|^2 + \left\| \frac{1}{2}V(t) \right\|^2 \leq \lambda^2.$$

Since $\|U(t)\| \leq \|V(t)\|$ and $\inf_{t \in \mathbb{T}} \|U(t)\| \geq 2\rho$, the above inequality yields

$$\left\| \frac{1}{2}(U_1(t) + U_2(t)) \right\| \leq \sqrt{\lambda^2 - \rho^2}$$

$$< \lambda.$$

Since U_1 and U_2 are solutions of (3.63), it follows that $\frac{1}{2}(U_1 + U_2)$ is also a solution of (3.63). Thus, $\frac{1}{2}(U_1 + U_2)$ is a solution of (3.63) with norm less than λ. This contradicts the fact that λ is least-value of solutions. Consequently, $U_1(t) = U_2(t)$ for each $t \in \mathbb{T}$.

Finally, for each $B \in H(A)$ and $G \in H(F)$, Lemma 3.29 guarantees that the unique minimum norm solution of (3.63) is almost periodic. This completes the proof. □

Theorem 3.31. *If SDE* (3.32) *admits an exponential dichotomy, then it has only one bounded solution* $U(t) = \theta$.

Proof. Let U be a solution of SDE (3.32). Then,

$$U(t) = e_A(t, 0)U(0) \quad \text{for } t \in \mathbb{T}.$$

For any sequence $\{\alpha_m\}_{m \in \mathbb{N}} \subset \mathcal{C}^*$, denote

$$A_m(t) = A(t + \alpha_m) \quad \text{and} \quad U_m(t) = U(t + \alpha_m).$$

The exponential dichotomy guarantees that U_m and A_m are bounded, that is, there exists constant $M > 0$ such that

$$\|U_m(t)\| \leq M \quad \text{and} \quad |A_m(t)| \leq M \quad \text{for all } t \in \mathbb{T}.$$

In view of $\Delta_g U_m(t) = A_m(t)U_m(t)$, we obtain

$$\|\Delta_g U_m(t)\| \leq M^2 \quad \text{for } m \in \mathbb{N}.$$

By Theorem 3.26, there exists a subsequence $\{\alpha_{m_i}\}$ of $\{\alpha_m\}_{m \in \mathbb{N}}$ such that $\{U_{m_i}\}$ converges uniformly on every compact subset of $\mathbb{T}$, that is,

$$\lim_{i \to \infty} U(t + \alpha_{m_i})$$

exists uniformly on $\mathbb{T}$. Consequently, by virtue of Theorem 3.20, $U(t)$ is almost periodic.

Since (3.32) admits an exponential dichotomy, it follows that $e_A(t,0) \to 0$ as $t \to \infty$. This implies that

$$\inf_{t \in \mathbb{T}} \|U(t)\| = 0.$$

Thereby, there exists a sequence $\{\alpha'_m\} \subset \mathbb{T}$ such that

$$\|U(\alpha'_m)\| \to 0 \quad \text{for } m \to \infty.$$

Since both A and U are almost periodic, we can extract a subsequence $\alpha := \{\alpha''_m\}$ of $\{\alpha'_m\}$ such that

$$T_\alpha A(t) = B(t),$$

$$T_\alpha U(t) = V(t),$$

$$T_{-\alpha} B(t) = A(t),$$

$$T_{-\alpha} V(t) = U(t)$$

uniformly on $\mathbb{T}$. Moreover, V is a solution of (3.62) and satisfies the initial condition

$$V(0) = T_\alpha U(0)$$

$$= \lim_{m \to \infty} U(\alpha_m)$$

$$= \theta.$$

Hence,

$$V(t) = e_B(t,0)V(0) = \theta.$$

Consequently, $U(t) = T_{-\alpha}V(t) = \theta$. This completes the proof. $\square$

Remark 3.8. From the proof of Theorem 3.31, it is clear that all equations in the hull of (3.32) have only one bounded solution $V(t) = \theta$ provided the SDE (3.32) admits an exponential dichotomy. In this case, (3.61) satisfies the Favard condition, thus the following Theorem 3.32 guarantees the existence and uniqueness of almost periodic solutions to (3.61).

In general, $(a-b)U \neq aU -_g bU$ for $a, b \in \mathbb{R}$ and $U \in K_c^n$, so we still need to assume the following equality:

$$e_A(t,0)(\mathcal{I} - P)e_{\ominus A^*}^*(s,0)F(s)$$

$$= e_A(t,0)e_{\ominus A^*}^*(s,0)F(s) -_g e_A(t,0)Pe_{\ominus A^*}^*(s,0)F(s) \quad \text{for } t, s \in \mathbb{T}.$$
$$(3.65)$$

Theorem 3.32. *Let A be an almost periodic $n \times n$-matrix-valued function. Suppose (3.32) admit exponential dichotomy on $\mathbb{T}$ with positive constants k, a and the projection P and $F \in \mathcal{C}_{\mathrm{rd}}(\mathbb{T}, K_c^n)$ be bounded. If the equality (3.65) is valid, then (3.61) has a unique bounded solution U satisfying*

$$
U(t) =
\begin{cases}
\displaystyle\int_0^t e_A(t,0) P e_{\ominus A^*}^*(s,0) F(s) \Delta_g s \\[2mm]
\quad -g \displaystyle\int_t^\infty e_A(t,0)(\mathcal{I}-P) e_{\ominus A^*}^*(s,0) F(s) \Delta_g s \quad \text{for } t \in \mathbb{T}^+, \\[4mm]
\displaystyle\int_{-\infty}^t e_A(t,0) P e_{\ominus A^*}^*(s,0) F(s) \Delta_g s \\[2mm]
\quad -g \displaystyle\int_t^0 e_A(t,0)(\mathcal{I}-P) e_{\ominus A^*}^*(s,0) F(s) \Delta_g s \quad \text{for } t \in \mathbb{T}^-.
\end{cases}
$$
(3.66)

Proof. It is sufficient to assume $t \in \mathbb{T}^+$. The argument is analogous for $t \in \mathbb{T}^-$. By means of Proposition 2.5 (i1)–(i3), we easily see that the right side of (3.66) is well-defined, and

$$
\int_0^t e_A(t,0) P e_{\ominus A^*}^*(s,0) F(s) \Delta_g s
$$
$$
-g \int_t^\infty e_A(t,0)(\mathcal{I}-P) e_{\ominus A^*}^*(s,0) F(s) \Delta_g s
$$
$$
= \int_0^\infty [e_A(t,0) P e_{\ominus A^*}^*(s,0) F(s)
$$
$$
-g\, e_A(t,0)(\mathcal{I}-P) e_{\ominus A^*}^*(s,0) F(s)] \Delta_g s
$$
$$
-g \int_t^\infty e_A(t,0) e_{\ominus A^*}^*(s,0) F(s) \Delta s
$$
$$
+ \int_0^t e_A(t,0)(\mathcal{I}-P) e_{\ominus A^*}^*(s,0) F(s) \Delta_g s.
$$

Next, we estimate the bound of the right side of (3.66). Let M be a positive constant such that

$$
\|F(s)\| \leq M \quad \text{for all } s \in \mathbb{T}.
$$

Then, by Definition 3.10 and Proposition 2.5(i6), we have

$$\left\| \int_0^t e_A(t,0)Pe^*_{\ominus A^*}(s,0)F(s)\Delta_g s \right\|$$

$$\leq \int_0^t \| e_A(t,0)Pe^*_{\ominus A^*}(s,0)F(s) \| \Delta s$$

$$= \int_0^t | e_A(t,0)Pe^*_{\ominus A^*}(s,0) | \| F(s) \| \Delta s$$

$$\leq k \int_0^t e_{\ominus a}(t,s) \| F(s) \| \Delta s$$

$$= k e_{\ominus a}(t,0) \int_0^t e_{\ominus a}(0,s) \| F(s) \| \Delta s$$

$$\leq k M \int_0^t e_{\ominus a}(0,s) \Delta s$$

$$\leq \frac{kM}{a}$$

and

$$\left\| \int_t^\infty e_A(t,0)(\mathcal{I} - P)e^*_{\ominus A^*}(s,0)F(s)\Delta_g s \right\|$$

$$\leq \int_t^\infty \| e_A(t,0)(\mathcal{I} - P)e^*_{\ominus A^*}(s,0)F(s) \| \Delta s$$

$$= \int_t^\infty | e_A(t,0)(\mathcal{I} - P)e^*_{\ominus A^*}(s,0) | \| F(s) \| \Delta s$$

$$\leq k \int_t^\infty e_{\ominus a}(s,t) \| F(s) \| \Delta s$$

$$\leq k M e_a(t,0) \int_t^\infty e_a(0,s) \Delta s$$

$$\leq \frac{kM}{a}.$$

From (3.66) and the above inequalities, it follows that

$$\| U(t) \| = D \left[\int_0^t e_A(t,0)Pe^*_{\ominus A^*}(s,0)F(s)\Delta_g s \right.$$

$$\left. -_g \int_t^\infty e_A(t,0)(\mathcal{I} - P)e^*_{\ominus A^*}(s,0)F(s)\Delta_g s, \{0\} \right]$$

$$= D\left[\int_0^t e_A(t,0)Pe^*_{\ominus A^*}(s,0)F(s)\Delta_g s,\right.$$

$$\left.\int_t^\infty e_A(t,0)(\mathcal{I}-P)e^*_{\ominus A^*}(s,0)F(s)\Delta_g s\right]$$

$$\leq \left\|\int_0^t e_A(t,0)Pe^*_{\ominus A^*}(s,0)F(s)\Delta_g s\right\|$$

$$+ \left\|\int_t^\infty e_A(t,0)(\mathcal{I}-P)e^*_{\ominus A^*}(s,0)F(s)\Delta_g s\right\|$$

$$\leq \frac{2kM}{a}.$$

This shows that U is uniformly bounded on $\mathbb{T}^+$. Now, it remains to show that U is a solution of (3.61). Differentiating (3.66) in the sense of Δ_g and taking into account the equality (3.65), together with Propositions 2.4, 2.5(15), and Theorem 1.9, we obtain

$$\Delta_g U(t) = e_A(\sigma(t),0)Pe^*_{\ominus A^*}(\sigma(t),0)F(t)$$

$$+ e_A(\sigma(t),0)(\mathcal{I}-P)e^*_{\ominus A^*}(\sigma(t),0)F(t)$$

$$+ \int_0^t A(t)e_A(t,0)Pe^*_{\ominus A^*}(s,0)F(s)\Delta_g s$$

$$-_g \int_t^\infty A(t)e_A(t,0)(\mathcal{I}-P)e^*_{\ominus A^*}(s,0)F(s)\Delta_g s$$

$$= F(t) + A(t)U(t).$$

Thus, U satisfies (3.61). Now, let U_1 and U_2 be two bounded solutions of (3.61). Then, by the same argument as in the proof of Theorem 3.30, we see that $U_1 -_g U_2$ is a bounded solution of (3.32). From Theorem 3.31, it follows that $U_1 = U_2$. This completes the proof. $\square$

Now, we are in a position to establish the existence and uniqueness of almost periodic solutions to nonlinear SDEs

$$\Delta_g U(t) = A(t)U(t) + F(t,U) \quad \text{for } t \in \mathbb{T}. \tag{3.67}$$

Define a distance on $AP(\mathbb{T}^+, K_c^n)$ as follows:

$$D_1[\Phi, \Psi] = \sup_{t \in \mathbb{T}^+} D[\Phi(t), \Psi(t)].$$

We find that $AP_1 =: (AP(\mathbb{T}^+, K_c^n), D_1)$ is a complete metric space.

Theorem 3.33. *Suppose that*

(i) *(3.32) admits an exponential dichotomy on $\mathbb{T}^+$ with positive constants k, a, a projection P and an almost periodic $n \times n$-matrix-valued function A.*

(ii) *$F \in AP_u(\mathbb{T}^+ \times K_c^n, K_c^n)$.*

(iii) *for each $\Phi \in AP_1$, the set-valued function $F_\Phi(s) = F(s, \Phi(s))$ satisfies the equality (3.65).*

(iv) *there exists a continuous nondecreasing function $\psi \colon \mathbb{R}^+ \to \mathbb{R}^+$ with $\psi(0) = 0$, $\psi(t) > 0$ for $t > 0$, and*

$$\lim_{t \to \infty} \psi(t) = \infty$$

such that

$$D[F(t, \Phi(t)), F(t, \Psi(t))] \leq \frac{a}{(2 + a\mu(t))k} \left[D_1[\Phi, \Psi] - \psi D_1(\Phi, \Psi) \right]$$

for all $t \in \mathbb{T}^+$ and $\Phi, \Psi \in AP_1$, where the constants k and a are given as in (i) and μ is the graininess function of $\mathbb{T}$.

Then, (3.67) has a unique almost periodic solution on $\mathbb{T}^+$.

Proof. For fixed $\Phi \in AP_1$, consider the linear set dynamic equation

$$\Delta_g U(t) = A(t)U(t) + F_\Phi(t) \quad \text{for } t \in \mathbb{T}^+. \tag{3.68}$$

According to Theorem 3.32 and Remark 3.8, SDE (3.68) has a unique almost periodic solution Ψ satisfying

$$\Psi(t) = \int_0^t e_A(t, 0) P e_{\ominus A^*}^*(s, 0) F_\Phi(s) \Delta_g s$$

$$-_g \int_t^\infty e_A(t, 0)(\mathcal{I} - P) e_{\ominus A^*}^*(s, 0) F_\Phi(s) \Delta_g s \quad \text{for } t \in \mathbb{T}^+. \tag{3.69}$$

Define a map $f \colon AP_1 \to AP_1$ by

$$f[\Phi](t) = \Psi(t) \quad \text{for } t \in \mathbb{T}^+,$$

where $\Psi(t)$ is given in (3.69). Let $P_1 = \mathcal{I} - P$. For $\Phi, \Psi \in AP_1$, in the virtue of properties of the Hausdorff distance, Proposition 2.5(17), and our hypotheses, we have

$$D[f[\Phi](t), f[\Psi](t)]$$

$$= D\left[\int_0^t e_A(t,0)Pe^*_{\ominus A^*}(s,0)F_\Phi(s)\Delta_g s\right.$$

$$- {}_g\int_t^\infty e_A(t,0)P_1 e^*_{\ominus A^*}(s,0)F_\Phi(s)\Delta_g s,$$

$$\int_0^t e_A(t,0)Pe^*_{\ominus A^*}(s,0)F_\Psi(s)\Delta_g s$$

$$\left. - {}_g\int_t^\infty e_A(t,0)P_1 e^*_{\ominus A^*}(s,0)F_\Psi(s)\Delta_g s\right]$$

$$\leq D\left[\int_0^t e_A(t,0)Pe^*_{\ominus A^*}(s,0)F_\Phi(s)\Delta_g s,\right.$$

$$\left.\int_0^t e_A(t,0)Pe^*_{\ominus A^*}(s,0)F_\Psi(s)\Delta_g s\right]$$

$$+ D\left[\int_t^\infty e_A(t,0)P_1 e^*_{\ominus A^*}(s,0)F_\Phi(s)\Delta_g s,\right.$$

$$\left.\int_t^\infty e_A(t,0)P_1 e^*_{\ominus A^*}(s,0)F_\Psi(s)\Delta_g s\right]$$

$$\leq \int_0^t D\left[e_A(t,0)Pe^*_{\ominus A^*}(s,0)F_\Phi(s), e_A(t,0)Pe^*_{\ominus A^*}(s,0)F_\Psi(s)\right]\Delta s$$

$$+ \int_t^\infty D[e_A(t,0)P_1 e^*_{\ominus A^*}(s,0)F_\Phi(s),$$

$$e_A(t,0)P_1 e^*_{\ominus A^*}(s,0)F_\Psi(s)]\Delta s$$

$$\leq \int_0^t |e_A(t,0)Pe^*_{\ominus A^*}(s,0)|D[F_\Phi(s), F_\Psi(s)]\Delta s$$

$$+ \int_t^\infty |e_A(t,0)P_1 e^*_{\ominus A^*}(s,0)|D[F_\Phi(s), F_\Psi(s)]\Delta s$$

$$\leq k \sup_{s\in\mathbb{T}^+} D[F_\Phi(s), F_\Psi(s)]\left[\int_0^t e_{\ominus a}(t,\sigma(s))\Delta s + \int_t^\infty e_{\ominus a}(\sigma(s),t)\Delta s\right]$$

$$\leq \frac{2 + a\mu(t)}{a} k \sup_{s\in\mathbb{T}^+} D[F_\Phi(s), F_\Psi(s)].$$

Keeping in mind the hypothesis (iv), we can write

$$D[f[\Phi](t), f[\Psi](t)] \leq D_1[\Phi, \Psi] - \psi D_1[\Phi, \Psi] \quad \text{for all } \in \mathbb{T}^+.$$

From the arbitrariness of t, it follows that

$$D_1[f[\Phi], f[\Psi]] \leq D_1[\Phi, \Psi] - \psi D_1[\Phi, \Psi].$$

Now, Theorem A.4 guarantees that f has a unique fixed point $\Phi \in AP_1$. This unique fixed point Φ is the unique almost periodic solution of (3.67) and can trivially be indicated by (3.69). This completes the proof. □

Remark 3.9. Theorem 3.33 involves the case of contractive condition. For example, if there exists a constant M with

$$M \in \left[0, \frac{a}{k(2 + a \sup_{t \in \mathbb{T}_+} \mu(t))}\right)$$

such that

$$D[F(t, \Phi(t)), F(t, \Psi)] \leq M D_1[\Phi, \Psi],$$

then f is contractive operator, where we assume $\sup_{t \in \mathbb{T}_+} \mu(t) < \infty$. Let

$$p = 1 - \frac{kM(2 + a \sup_{t \in \mathbb{T}_+} \mu(t))}{a}$$

and $\psi(t) = pt$. Thus, f is weakly contractive.

The following examples further illustrate the applicability of the results involved in this section.

Example 3.1. Let

$$A = \begin{pmatrix} -\lambda & 0 \\ 0 & -\lambda \end{pmatrix},$$

where $\lambda > 0$ such that $-\lambda \in \mathcal{R}_1^+$. Take $F(t) = f(t)C$, where

$$f(t) = \begin{pmatrix} \sin \sqrt{3}t \\ \cos \sqrt{2}t \end{pmatrix}$$

for $t \in \mathbb{T}$, and C is a bounded subset of $\mathbb{R}$. Then, (3.61) has a unique almost periodic solution C given by

$$
U(t) = \begin{cases}
C \displaystyle\int_0^t e_{-\lambda}(t,s) f(s) \mathcal{I}_2 \Delta_g s & \text{for } t \in \mathbb{T}^+ \\[3mm]
C \displaystyle\int_{-\infty}^t e_{-\lambda}(t,s) f(s) \mathcal{I}_2 \Delta_g s & \text{for } t \in \mathbb{T}^-
\end{cases}
\tag{3.70}
$$

with

$$
\mathcal{I}_2 = \begin{pmatrix} 1 & 0 \\ 0 & 1 \end{pmatrix}.
$$

Proof. Obviously, A and F satisfy the conditions of Theorem 3.32. From Theorem 3.16, it follows that the homogeneous equation corresponding to (3.61) admits an exponential dichotomy with the constants $k = a = 1$, the projection $P = \mathcal{I}_2$ and the fundamental matrix $e_A(t,0) = e_{-\lambda}(t,0)\mathcal{I}_2$. The equality (3.65) is naturally satisfied. Now, Theorem 3.32 and Remark 3.8 guarantees that the existence of the unique almost periodic solution U to (3.61) and the representation of U in the required form (3.70) is immediately obtained by (3.66). $\square$

Example 3.2. Let $\mathbb{I}$ be a class of all closed bounded intervals in $\mathbb{R}$. Consider an interval-valued dynamic equation

$$
\Delta_g U(t) = -p(t) U(t) + F(t, U) \quad \text{for } t \in \mathbb{T}^+, \tag{3.71}
$$

where $U(t) = [u^-(t), u^+(t)]$ and $F(t, U) = [f^-(t, U), f^+(t, U)]$. Then, (3.71) has a unique almost periodic solution on $\mathbb{T}^+$ provided the following hypotheses hold.

(I) $p \in AP(\mathbb{T}^+, \mathbb{R}^+)$, $\inf_{t \in \mathbb{T}^+} p(t) > 0$, and $-p \in \mathcal{R}_1^+$.

(II) $f^-, f^+ \in AP_u(\mathbb{T}^+ \times \mathbb{I}, \mathbb{R})$.

(III) If $U, V \in AP(\mathbb{T}^+, \mathbb{I})$, then, for each $t \in \mathbb{T}^+$,

$$
\max \left\{ |f^-(t, V(t)) - f^-(t, U(t))|, |f^+(t, V(t)) - f^+(t, U(t))| \right\}
$$
$$
\leq \frac{[D_1[U, V] - \ln(1 + D_1[U, V])]}{(2 + \mu(t))}.
$$

Proof. Again, Theorem 3.16 guarantees that the homogeneous equation

$$\Delta_g U(t) = p(t) U(t)$$

admits an exponential dichotomy with constants $k = a = 1$, the projection $P = \mathcal{I}$ and the fundamental matrix $e_{-p}(t, 0)$, that is, the condition (i) of Theorem 3.33 is satisfied. For $(t_1, U_1), (t_2, U_2) \in \mathbb{T}^+ \times \mathbb{I}$, set

$$f_1^- = f^-(t_1, U_1), \quad f_2^- = f^-(t_2, U_2) \quad \text{and} \quad f_1^+ = f^+(t_1, U_1),$$
$$f_2^+ = f^+(t_2, U_2).$$

By analogues to classical case, Theorem A.6, we have

$$D[F(t_1, U_1), F(t_2, U_2)]$$
$$= \max \left\{ | \min\{f_1^- - f_2^-, f_1^+ - f_2^+\} |, | \max\{f_1^- - f_2^-, f_1^+ - f_2^+\} | \right\}.$$

This easily infers the continuity of F. Since both f^- and f^+ are almost periodic single-valued functions, similar to the case of $\mathbb{T} = \mathbb{R}$, one can obtain that, for any $\varepsilon > 0$,

$$\mathscr{E}(\varepsilon, f^-) \cap \mathscr{E}(\varepsilon, f^+) \neq \emptyset.$$

Therefore, there exists $\omega \in \mathcal{C}^*$ such that

$$\left| \min\{f^-(t + \omega, U) - f^-(t, U), f^+(t + \omega, U) - f^+(t, U)\} \right| < \varepsilon,$$
$$\left| \max\{f^-(t + \omega, U) - f^-(t, U), f^+(t + \omega, U) - f^+(t, U)\} \right| < \varepsilon$$

for any $U \in \mathbb{I}$. Thus,

$$D[F(t + \omega, U), F(t, U)] < \varepsilon$$

for any $U \in \mathbb{I}$. This implies $F \in AP_u(\mathbb{T}^+ \times \mathbb{I}, \mathbb{I})$. Consequently, F satisfies (ii) of Theorem 3.33. Further, (iii) of Theorem 3.33 is trivially valid. Finally, (iv) of Theorem 3.33 is also valid, from (III) by taking $\psi(t) = \ln(1 + t)$. Consequently, Theorem 3.33 reaches our desired conclusion. $\square$

Example 3.3. Consider a special case for set dynamic equation (3.71) by taking $p(t) \equiv c > 0$, and

$$f^-(t, U) = r(U) \sin t, \ f^+(t, U) = 2r(U) \sin t \quad \text{for } U \in AP(\mathbb{T}^+, \mathbb{I}),$$

where $r \colon \mathbb{I} \to \mathbb{R}$ is bounded and satisfies

$$|r(V) - r(U)| \leq \frac{(D[V, U])^2}{4(1 + D[V, U])}.$$

In this sense, (I) and (II) of Example 3.2 are satisfied. Further,

$$|f^+(t, V) - f^+(t, U)| = 2|\sin t||r(V) - r(U)|$$

$$\leq \frac{(D[V(t), U(t)])^2}{2(1 + D[V(t), U(t)])}$$

$$\leq \frac{1}{2}(D_1[V, U] - \ln(1 + D_1[V, U])),$$

that is, (III) of Example 3.2 is satisfied. Now, employing the definition of integration of an interval-valued function given in Stefanini and Bede (2009) to time scale case, (3.69) yields the unique almost periodic solution Ψ of (3.71) represented by

$$\Psi(t) = \int_0^t e_p(t, s)F(s, \Psi)\Delta_g s$$

$$= r(\Psi(t))\left[\int_0^t e_c(t, s)\sin s\Delta s, 2\int_0^t e_c(t, s)\sin s\Delta s\right]$$

$$= [1, 2]r(\Psi(t))\int_0^t e_c(t, s)\sin s\Delta s.$$

In particular, if $\mathbb{T} = \mathbb{R}$, then $e_c(t, s) = e^{c(t-s)}$ and

$$\Psi(t) = [1, 2]\frac{r(\Psi(t))}{1 + c^2}(1 - \cos t - c\sin t).$$

If $\mathbb{T} = \mathbb{Z}$, then $e_c(t, s) = (1 + c)^{t-s}$ and

$$\Psi(t) = [1, 2](1 + c)^t r(\Psi(t))\sum_{i=1}^t \frac{\sin i}{(1 + c)^i}.$$

By applying the modified fuzzy theory discussed in the previous section, Section 3.5.2, Theorem 3.32 can be translated to the fuzzy version

$$\Delta_g Y(t) = p(t)Y(t) + B(t) \quad \text{for } t \in \mathbb{T}^+ \tag{3.72}$$

of the first-order linear set dynamic equation (3.61). The existence of local solutions for the initial problem of fuzzy dynamic equation (3.72) on $\mathbb{R}$ has been discussed in Khastan *et al.* (2011). As an application of Theorem 3.32, we now establish the existence of almost periodic solutions to (3.72) on time scale $\mathbb{T}$.

Theorem 3.34. *Assume that $p \in AP(\mathbb{T}^+, \mathbb{R}) \cap \mathcal{R}_1^+$ is such that*

$$\sup_{t \in \mathbb{T}} p(t) < 0$$

and $B \in AP(\mathbb{T}, \mathbb{T}_f)$. Then, the fuzzy dynamic equation (3.72) has a unique almost periodic solution on $(AP(\mathbb{T}^+, \mathbb{T}_f), D)$ satisfying

$$Y(t) = \int_0^t e_p(t, s) B(s) \Delta_g s. \tag{3.73}$$

Proof. It suffices to check that

$$\Delta_g Y(t) = p(t)Y(t)$$

admits exponential dichotomy on $\mathbb{T}$ and the equality (3.65) is valid. To this end, take

$$k = 1, \quad a = -\sup_{t \in \mathbb{T}} p(t) > 0, \quad \text{and} \quad P = \mathcal{I}.$$

Then, we have

$$0 < 1 + p(\tau)\mu(\tau)$$
$$\leq 1 - a\mu(\tau)$$
$$< 1 - \frac{a\mu(\tau)}{1 + a\mu(\tau)}$$

That is,

$$\frac{1}{1 + a\mu(\tau)} > 0 \quad \text{for } \tau \in [s, t]_{\mathbb{T}},$$

which implies that

$$e_p(t, s) = \exp\left[\int_s^t \frac{\text{Log}(1 + p(\tau)\mu(\tau))}{\mu(\tau)} \Delta\tau\right]$$

$$\leq \exp\left[\int_{\sigma(s)}^t \frac{\text{Log}(1 + p(\tau)\mu(\tau))}{\mu(\tau)} \Delta\tau\right]$$

$$\leq \exp\left[\int_{\sigma(s)}^t \frac{\text{Log}\left[\frac{1}{(1+a\mu(\tau))}\right]}{\mu(\tau)} \Delta\tau\right]$$

$$= e_{\ominus a}(t, \sigma(s))$$

for $s, t \in \mathbb{T}$ with $t \geq \sigma(s)$. Thus, we have

$$|e_p(t, 0)Pe_{\ominus p}(s, 0)| = e_p(t, s)$$

$$\leq e_{\ominus a}(t, \sigma(s)), \quad s, t \in \mathbb{T}, \quad t \geq \sigma(s).$$

This deduces the exponential dichotomy. The equality (3.65) is obviously valid. Now, Theorem 3.32, in view of Remark 3.8, guarantees the existence and uniqueness of the almost periodic solution Y to (3.72). From (3.66), this unique solution is given by

$$Y(t) = \int_0^t e_p(t, 0)Ie_{\ominus p}(s, 0)B(s)\Delta_g s$$

$$= \int_0^t e_p(t, s)B(s)\Delta_g s.$$

This completes the proof. □

Remark 3.10. If $\mathbb{T} = \mathbb{R}$, then

$$e_p(t, s) = e^{\int_s^t p(\tau)\Delta\tau}$$

$$= e^{\int_0^t p(\tau)\Delta\tau - \int_0^s p(\tau)\Delta\tau}$$

and

$$Y(t) = e^{\int_0^t p(\tau)\Delta\tau}\int_0^t B(s)e^{-\int_0^s p(\tau)\Delta\tau}\Delta_g s.$$

If $\mathbb{T} = \mathbb{Z}$ and $p(t) = p < 0$, then $e_p(t, s) = (1 + p)^{t-s}$ and

$$Y(t) = (1 + p)^t \int_0^t \frac{1}{(1 + p)^s} B(s) \Delta_g s$$

$$= (1 + p)^t \sum_{i=0}^t \frac{1}{(1 + p)^i} B(i).$$

Example 3.4. Let

$$p(t) = -\frac{1}{4}(2 - \cos t) \quad \text{and} \quad B(t) = \gamma \sin 2t,$$

where $\gamma \in \mathbb{T}_F$ with

$$[\gamma]^r = [r - 1, 1 - r] \cap \mathbb{T} \quad \text{for } r \in [0, 1].$$

If $\mathbb{T} = \mathbb{R}$, then $\mu(t) = 0$, $p \in \mathcal{R}_1^+$, and

$$e_p(t, s) = e^{\frac{1}{4}(2s - 2t + \sin t - \sin s)}.$$

As a conclusion, the fuzzy dynamic equation (3.72) has a unique almost periodic solution

$$Y(t) = e^{\frac{1}{4}(-2t + \sin t)} \gamma \int_0^t \sin 2s \, e^{-\frac{1}{4}(2s - \sin s)} \Delta_g s.$$

Moreover, taking $p(t) = -1$ and $\mathbb{T} = \frac{1}{2}\mathbb{Z}$, we have $\mu(t) = \frac{1}{2}$, $p \in \mathcal{R}_1^+$, and

$$e_p(t, s) = \left(1 + \frac{1}{2}p\right)^{2(t-s)}.$$

Therefore, the solution of (3.72) has the form

$$Y(t) = \left(1 + \frac{1}{2}p\right)^{2t} \gamma \sum_{i=0}^t \frac{1}{(1 + \frac{1}{2}p)^{2i}} \sin 2i$$

$$= 4^t \gamma \sum_{i=0}^t \frac{\sin 2i}{4^i}.$$

Chapter 4

Stability of Set Dynamic Equations

4.1 Background

Stability of solutions for SDEs on time scales has recently received much attention and is undergoing rapid development (see Hong, 2010; Hong *et al.*, 2014; Jia *et al.*, 2022; Li and Hong, 2011). In this chapter, by defining appropriate Lyapunov-like functions and formulating certain inequalities on the functions under consideration, we explore the stability for the solutions of SDEs on time scales. Moreover, we employ these results to investigate the exponential stability, exponentially asymptotic stability, uniform stability, and uniformly exponentially asymptotic stability for the trivial solution of SDEs.

Notions of stability for the solutions of set dynamic equations on time scales are considered by using Lyapunov-like functions. Criteria for equi-stability, equi-asymptotic stability, uniform stability, and uniform asymptotic stability are developed. The exponential stability, exponentially asymptotic stability, uniform exponential stability, and uniformly exponentially asymptotic stability for the trivial solution of set dynamic equations on time scales by using Lyapunov-like functions are considered. We present a new definition for exponential stability of solutions, including H-exponential stability, H-exponentially asymptotic stability, H-uniformly exponential stability, and H-uniformly exponentially asymptotic stability for a class of set dynamic equations on time scales. Employing Lyapunov-type functions on time scales, we provide sufficient conditions for

147

the exponential stability of the trivial solution for such set dynamic equations.

4.2 Method of Exponential Dichotomy

Continuing Section 3.4.1, in this section, we consider the stability of ISDE (3.26) by employing the exponential dichotomy. We assume that $F(t, \{0\}) \equiv \{0\}$ and

$$J_k(\{0\}) + H_k = \{0\} \quad \text{for } t \in \mathbb{T}^+ \text{ and } k \in \mathbb{N}_0.$$

Moreover, by $F \in \mathrm{Lip}(h)$, we mean that $F \in PC[\mathbb{T}^+ \times \mathcal{C}_{\mathrm{rd}}, K_k^n]$, and

$$D[F(t, U), F(t, V)] \le h(t)D[U, V] \quad \text{for all } t \in \mathbb{T}^+, \ U, V \in PC^1, \tag{4.1}$$

where $h \colon \mathbb{T}^+ \to \mathbb{R}^+$ satisfies

$$k(1 + a\mu(t))h(t) < a$$

and

$$\int_0^\infty k(1 + a\mu(s))h(s)\Delta s < \infty$$

with the constants k and a given as in Definition 3.10 and μ, a graininess function for $\mathbb{T}$. For the sake of convenience, we assume the projection $P \equiv \mathcal{I}$.

Under the assumption of $F \in \mathrm{Lip}(h)$, employing the procedure used in the proof of Theorem 3.19, we obtain that the impulsive set dynamic equation

$$\begin{cases} \Delta_g U(t) = A(t)U(t) + F(t, U) & \text{for } t \ne t_k, \ k \in \mathbb{N}_0 \\ U_{t_k^+} = \Phi_k(t) = J_k(U(t_k)) + H_k & \text{for } t = t_k, \ k \in \mathbb{N}_0 \end{cases} \tag{4.2}$$

has a solution

$$V_k(t) = V_k(t_k, \Phi_k)(t)$$

$$= e_A(t, t_k^+)\Phi_k + \int_{t_k}^t e_A(t, \sigma(t))F(\tau, V_k)\Delta\tau \quad \text{for } t \in (t_k, t_{k+1}]_{\mathbb{T}^+}$$

for $k \in \mathbb{N}_0$. Additionally, define $V_0(t_0) = U_0$, $V_k(t_k) = V_{k-1}(t_k)$ for $k \in \mathbb{N}$. We now obtain a solution $U(t_0, U_0)$ of ISDE (3.45) on $\mathbb{T}^+$

which is left continuous on $(t_k, t_{k+1}]_{\mathbb{T}+}$ and defined by

$$U(t_0, U_0)(t) = \begin{cases} \mathscr{U}_0(t, t_0, U_0) & \text{for } t \in [0, t_0]_{\mathbb{T}+} \\ V_0(t) & \text{for } t \in (t_0, t_1]_{\mathbb{T}+} \\ \vdots & \vdots \\ V_k(t) & \text{for } t \in (t_k, t_{k+1}]_{\mathbb{T}+} \\ \vdots & \vdots \end{cases}$$

On the other hand, with the additional assumption that $P \equiv \mathcal{I}$, from (3.46), we obtain a bounded solution of ISDE (3.45) given by

$$\overline{U}(t) = \begin{cases} e_A^0(t)U_0 + \displaystyle\int_{t_0}^t e_A(t, \sigma(\tau))F(\tau, \overline{U})\Delta_g\tau & \text{for } t \in [0, t_0]_{\mathbb{T}+} \\ \displaystyle\int_{t_0}^t e_A^0(t)e_{\ominus A^*}^0(\sigma(\tau))F(\tau, \overline{U})\Delta_g\tau \\ \quad + \displaystyle\sum_{t_j < t} e_A^0(t)e_{\ominus A^*}^0(t_j^+)H_j & \text{for } t \in (t_0, \infty)_{\mathbb{T}+}. \end{cases}$$

Since

$$e_A^0(t)e_{\ominus A^*}^0(\sigma(\tau)) = e_A(t, t_0)e_A(t_0, \sigma(\tau))$$
$$= e_A(t, \sigma(\tau)),$$

from $\overline{U}$ and V_k, we can write

$$V_k(t) -_g \overline{U}(t) = e_A(t, t_k^+)\Phi_k + \int_{t_k}^t e_A(t, \sigma(\tau))F(\tau, V_k)\Delta_g\tau$$

$$-_g \int_{t_0}^t e_A^0(t)e_{\ominus A^*}^0(\sigma(\tau))F(\tau, \overline{U})\Delta_g\tau$$

$$+ \sum_{t_j < t} e_A^0(t)e_{\ominus A^*}^0(t_j^+)H_j$$

$$= e_A(t, t_k^+)\Phi_k$$

$$+ \int_{t_k}^t e_A(t, \sigma(\tau))\left(F(\tau, V_k) -_g F(\tau, \overline{U})\right)\Delta_g\tau$$

$$+ \sum_{t_j < t} e_A^0(t)e_{\ominus A^*}^0(t_j^+)H_j.$$

Thus, we have the following attractive result.

Theorem 4.1. *Suppose that (3.32) admits an exponential dichotomy on $[t_0, \infty)_{\mathbb{T}^+}$ with positive constants k, a, the projection $P \equiv \mathcal{I}$, and the $n \times n$-matrix-valued function A which satisfy the hypothesis (H2), and $F \in Lip(h)$. Further, we assume that*

(i) *for each $\Phi \in PC^1$, the set-valued function $F_\Phi(s) = F(s, \Phi(s))$ satisfies the hypothesis (H1),*

(ii) *there exists $l > 0$ such that $\|\Phi_k\| \leq l$ for $k \in \mathbb{N}_0$.*
Then,

$$\lim_{t \to \infty} D[U(t_0, U_0)(t), \overline{U}(t)] = 0.$$

Proof. Let $t \in (t_k, t_{k+1}]_{\mathbb{T}^+}$ for some $k \in \mathbb{N}$. Note that

$$e_A(t, t_k^+) = e_A(t, \sigma(t_k)),$$

and combining the exponential dichotomy, and the assumption (ii), we have

$$D\left[U(t_0, U_0)(t), \overline{U}(t)\right] = D\left[V_k(t), \overline{U}(t)\right] = \left\| V_k(t) -_g \overline{U}(t) \right\|$$

$$\leq \left| e_A(t, t_k^+) \right| \|\Phi_k\| + \int_{t_k}^t \left| e_A(t, \sigma(s)) \right| \left\| \left[F(s, V_k) -_g F(s, \overline{U}) \right] \right\| \Delta s$$

$$+ \sum_{t_j < t} \left| e_A^0(t) e_{\ominus A^*}^0(t_j^+) \right| \|H_j\|$$

$$\leq k e_{\ominus a}(t, \sigma(t_0)) \|\Phi_k\| + \int_{t_k}^t k e_{\ominus a}(t, \sigma(s)) h(s) D[V_k(s), \overline{U}(s)] \Delta s$$

$$+ k \sum_{t_j < t} e_{\ominus a}(t, \sigma(t_j)) \|H_j\|$$

and therefore, in view of (H2),

$$e_a(t, 0) D\left[U(t_0, U_0)(t), \overline{U}(t)\right]$$

$$\leq k e_a(\sigma(t_0), 0) \|\Phi_k\| + k \|\mathscr{H}\|_\infty \sum_{t_j < t} e_a(\sigma(t_j), 0)$$

$$+ \int_{t_k}^t k e_a(\sigma(s), 0) h(s) D\left[U(t_0, U_0)(s), \overline{U}(s)\right] \Delta s$$

$$\leq ke_a(\sigma(t_0),0)\,\|\Phi_k\| + \frac{k}{1-\chi}\,\|\mathscr{H}\|_\infty$$

$$+ \int_{t_k}^t k(1+a\mu(s))h(s)e_a(\sigma(s),0) \times D\left[U(t_0,U_0)(s),\overline{U}(s)\right]\Delta s.$$

In view of Gronwall's inequality, given in Theorem 1.13, we obtain

$$e_a(t,0)D\left[U(t_0,U_0)(t),\overline{U}(t)\right] \leq l_k + l_k \int_{t_k}^t e_p(t,\sigma(s))p(s)\Delta s$$

$$\leq l_k + l_k \int_{t_k}^t p(s)e_p(t,0)\Delta s,$$

where $p(s) = k(1+a\mu(s))h(s)$ and

$$l_k = ke_a(\sigma(t_0),0)\,\|\Phi_k\| + \frac{k}{1-\chi}\,\|\mathscr{H}\|_\infty.$$

From this, it follows that

$$D\left[U(t_0,U_0)(t),\overline{U}(t)\right] \leq l_k e_{\ominus a}(t,0) + l_k e_{p\ominus a}(t,0)\int_{t_k}^t p(s)\Delta s,$$

which guarantees that

$$D\left[U(t_0,U_0)(t),\overline{U}(t)\right] \to 0 \quad \text{as} \quad t \to \infty.$$

This completes the proof. $\square$

We are now in a position to formulate the stability criteria for the trivial solution of ISDE (3.26). Let us first define the stability of trivial solution.

Definition 4.1. Let $U(t) = U(t_0,U_0)(t)$ be any solution of ISDE (3.26). Then the trivial solution $U(t) \equiv \theta$ is said to be stable provided for each $\varepsilon > 0$ and $t_0 \in \mathbb{T}^+$, there exists $\delta = \delta(t_0,\varepsilon) > 0$ such that $D[U_0,\theta] < \delta$ implies that

$$D[U(t),\theta] < \varepsilon$$

for all $t \in \mathbb{T}^+$.

Theorem 4.2. *Under the assumptions of Theorem 4.1, the trivial solution of ISDE (3.26) is stable.*

Proof. From the above arguments, we see that any solution of ISDE (3.26) can be indicated by

$$U(t) = \begin{cases} V(t) & \text{for } t \in [0, t_0]_{\mathbb{T}+} \\ W(t) & \text{for } t \in (t_0, \infty)_{\mathbb{T}+}, \end{cases}$$

where

$$V(t) = e_A^0(t)U_0 + \int_{t_0}^t e_A(t, \sigma(s))F(s, U)\Delta_g s \quad \text{for } t \in [0, t_0]_{\mathbb{T}+}$$

and

$$W(t) = \int_{t_0}^t e_A^0(t)e_{\ominus A^*}^0(\sigma(s))F(s, U)\Delta_g s \quad \text{for } t \in (t_0, \infty)_{\mathbb{T}+}.$$

We first consider the case of $t \in [0, t_0]_{\mathbb{T}+}$. Similar to the proof of Theorem 4.1, we obtain

$$e_a(t, 0)D[U(t), \theta] \le l_0 + l_0 e_p(t, 0) \int_0^{t_0} p(s)\Delta s,$$

where $p(s) = k(1 + a\mu(s))h(s)$, $l_0 = ke_a(\sigma(t_0), 0)D(U_0, \theta)$. For any given $\varepsilon > 0$, choose

$$\delta_1 = \varepsilon \left[ke_a(\sigma(t_0), 0) \left(1 + e_a(t_0, 0) \int_0^{t_0} p(s)\Delta s \right) \right]^{-1},$$

and then we have

$$D[U(t), \theta] < \varepsilon$$

whenever

$$D[U_0, \theta] < \delta_1.$$

If $t \in (t_0, \infty)_{\mathbb{T}+}$, then by the exponential dichotomy, we have

$$D[U(t), \theta] \le D[W(t), \theta] + e_a(0, t)D[U_0, \theta]$$

$$\le \int_{t_0}^t \left| e_{\ominus A^*}^0(\sigma(s)) \right| D[F(s, U), F(t, \theta)]\Delta s + e_a(0, t)D[U_0, \theta]$$

$$\le \int_{t_0}^t ke_{\ominus a}(t, \sigma(s))h(s)D[U(s), \theta]\Delta s + e_a(0, t)D[U_0, \theta]$$

and therefore

$$e_a(t,0)D\left[U(t),\theta\right] \leq \int_{t_0}^{t} ke_a(\sigma(s),0)h(s)D[U(s),\theta]\Delta s + D[U_0,\theta]$$

$$= \int_{t_0}^{t} p(s)e_a(s,0)D[U(s),\theta]\Delta s + D[U_0,\theta].$$

In view of Gronwall's inequality, given in Theorem 1.13, we obtain

$$e_a(t,0)D\left[U(t),\theta\right] \leq D[U_0,\theta]\int_{t_0}^{t} p(s)e_p(t,\sigma(s))\Delta s + D[U_0,\theta],$$

which yields that

$$D\left[U(t),\theta\right] \leq D[U_0,\theta]\left(1 + \int_{t_0}^{\infty} p(s)\Delta s\right).$$

Taking $\delta_2 = \varepsilon\left(1 + \int_{t_0}^{\infty} p(s)\Delta s\right)^{-1}$, we have

$$D\left[U(t),\theta\right] < \varepsilon$$

provided

$$D\left[U_0,\theta\right] < \delta_2.$$

Let $\delta = \min\{\delta_1, \delta_2\}$. Then,

$$D\left[U_0,\theta\right] < \delta$$

implies that

$$D\left[U(t),\theta\right] < \varepsilon \quad \text{for } t \in \mathbb{T}^+.$$

This completes the proof. $\qquad\square$

4.3 Equi-stability Criteria

4.3.1 *Comparison theorem*

In this section, we mainly adopt the Lyapunov functional approach to investigate the qualitative behaviour, such as the equistability, equiasymptotic stability, uniform stability, and uniform asymptotic stability, of the solutions of the following set dynamic initial value problem (SDIVP)

$$\Delta_g X(t) = F(t, X), \quad X(t_0) = X_0 \in K_c^n, \quad t \in \mathbb{T}, \qquad (4.3)$$

where $F \in \mathcal{C}_{\mathrm{rd}}(\mathbb{T}, K_c^1)$. To this end, we shall first formulate a comparison theorem for the solutions of SDIVP (4.3). As an application of the comparison theorem, we also prove the global existence result. In what follows, we assume that $\mathbb{T}$ is a time scale with $t_0 \geq 0$ as the minimal element and has no maximal element. From now on, unless otherwise mentioned, $U(t, t_0, U_0)$ stands for the solution of SDIVP (4.3) on $J_{\mathbb{T}}$ corresponding the initial value (t_0, U_0).

Theorem 4.3 (Comparison result). *Let* $\Omega = J_{\mathbb{T}} \times \{U \in K_c^1 : D[U, U_0] \leq b\}$ *with* $J_{\mathbb{T}} = [t_0, t_0 + a]_{\mathbb{T}}$ *and* $a, b > 0$. *Assume that* $F \in \mathcal{C}_{\mathrm{rd}}(\Omega, K_c^1)$ *is such that*

$$D[F(t, U(t)), F(t, V(t))] \leq G(t, D[U(t), V(t)]),$$

where $U(t) = U(t, t_0, U_0)$, $V(t) = V(t, t_0, U_0) \in K_c^1$ *for* $t \in \mathbb{T}$, $G \in \mathcal{C}_{\mathrm{rd}}(\Omega_0, \mathbb{R}^+)$ *with*

$$\Omega_0 = J_{\mathbb{T}} \times \{w \in \mathbb{R} : |w - w_0| \leq b\}$$

and $G(t, w)$ *is nondecreasing in* w *for each* $t \in J_{\mathbb{T}}$. *Then there exists the maximal solution* $r(t, t_0, w_0)$ *of dynamic initial value problem*

$$w^\Delta(t) = G(t, w), \quad w(t_0) = w_0 \geq 0, \quad \text{for } t \in J_{\mathbb{T}} \qquad (4.4)$$

such that

$$D[U(t), V(t)] \leq r(t, t_0, w_0) \quad \text{for } t \in J_{\mathbb{T}},$$

provided that $D[U_0, V_0] \leq w_0$.

Proof. Since U and V are solutions of SDIVP (4.3), the g-differences $U(s) -_g U(t)$ and $V(s) -_g V(t)$ exist for small $s - t > 0$. Set

$$m(t) = D[U(t), V(t)] \quad \text{for } t \in \mathbb{T}.$$

As an application of the properties of Hausdorff metric, we obtain the estimation

$$
\begin{aligned}
m(s) - m(t) &= D[U(s), V(s)] - D[U(t), V(t)] \\
&\leq D[U(s), U(t) + (s-t)F(t, U)] \\
&\quad + D[U(t) + (s-t)F(t, U), V(t) + (s-t)F(t, V)] \\
&\quad + D[V(t) + (s-t)F(t, V), V(s)] - D[U(t), V(t)] \\
&\leq D[U(s), U(t) + (s-t)F(t, U)] \\
&\quad + D[V(t) + (s-t)F(t, V), V(s)] \\
&\quad + (s-t)D[F(t, U), F(t, V)] \\
&= D[U(s) -_g U(t), (s-t)F(t, U)] \\
&\quad + D[(s-t)F(t, V), V(s) -_g V(t)] \\
&\quad + (s-t)D[F(t, U), F(t, V)].
\end{aligned}
$$

This implies that

$$
\begin{aligned}
\frac{m(s) - m(t)}{s - t} &\leq D\left[\frac{U(s) -_g U(t)}{s - t}, F(t, U)\right] \\
&\quad + D\left[F(t, V), \frac{V(s) -_g V(t)}{s - t}\right] + D[F(t, U), F(t, V)].
\end{aligned}
\tag{4.5}
$$

If t is right-dense, then taking $\overline{\lim}$ sup as $s \to t^+$, (4.5), together with Proposition 2.3(III), yields

$$
\begin{aligned}
m_+^\Delta(t) = \overline{\lim}_{s \to t^+} \sup \frac{m(s) - m(t)}{s - t} \\
\leq D[\Delta_g U(t), F(t, U)] + D[\Delta_g V(t), F(t, V)] \\
+ D[F(t, U), F(t, V)] \\
= D[F(t, U), F(t, V)],
\end{aligned}
$$

where $m_+^\Delta(t)$ is the delta right-derivative of m at $t \in \mathbb{T}$. On the other hand, if t is a right-scattered, then take $s = \sigma(t)$ in (4.5).

From Proposition 2.3 (III), it follows that

$$m^\Delta(t) = \frac{m(\sigma(t)) - m(t)}{\mu(t)}$$

$$\leq D\left[\frac{U(\sigma(t)) -_g U(t)}{\sigma(t) - t}, F(t, U)\right]$$

$$+ D\left[F(t, V), \frac{V(\sigma(t)) -_g V(t)}{\sigma(t) - t}\right] + D[F(t, U), F(t, V)]$$

$$= D[\Delta_g U(t), F(t, U)] + D[\Delta_g V(t), F(t, V)]$$

$$+ D[F(t, U), F(t, V)]$$

$$= D[F(t, U), F(t, V)].$$

In addition, the fact that $D[U_0, V_0] \leq w_0$ implies that $m(t_0) \leq w_0$. Consequently, the comparison theorem, Theorem A.7 gives

$$D[U(t), V(t)] \leq r(t, t_0.w_0), \quad t \in J_\mathbb{T}.$$

This completes the proof.   □

Remark 4.1. Assume F satisfies the hypotheses of Theorem 4.3 and $\|F(t, U)\| \leq M_0$ on Ω. Also, assume that $G \in \mathcal{C}_{rd}(J_\mathbb{T} \times [0, 2b], \mathbb{R}^+)$ is such that $G(t, w) \leq M_1$ on $[0, 2b]$, $G(t, 0) \equiv 0$, $G(t, w)$ is increasing in w for each $t \in J_\mathbb{T}$ and $w(t) \equiv 0$ is the only solution of dynamic initial value problem

$$w^\Delta(t) = g(t, w), \quad w(t_0) = 0.$$

Then the successive approximations defined by

$$U_{n+1}(t) = U_0 + \int_{t_0}^{t} F(s, U_n(s))\Delta_g s, \quad n \in \mathbb{N}_0$$

exist as continuous functions on $I_\mathbb{T} = [t_0, t_0 + \alpha]_\mathbb{T}$, where

$$\alpha = \min\left(a, \frac{b}{M}\right), \quad M = \max\{M_0, M_1\},$$

and the sequence $\{u_n\}$ converges uniformly to the unique solution $U(t) = U(t, t_0, U_0)$ of SDIVP (4.3) on $I_\mathbb{T}$.

The proof of Remark 4.1 exhibits the idea of comparison result.

Theorem 4.4. *Assume that*

(s1) $F \in \mathcal{C}_{\mathrm{rd}}(\mathbb{T} \times K_c^1, K_c^1)$ *satisfies*

$$\|F(t, \Phi)\| \leq G(t, \|\Phi\|)$$

for each $(t, \Phi) \in \mathbb{T} \times K_c^1$, *where* $G \in \mathcal{C}_{\mathrm{rd}}(\mathbb{T} \times \mathbb{R}^+, \mathbb{R}^+)$ *and* $G(t, w)$ *is nondecreasing in* w *for each* $t \in \mathbb{T}$.

(s2) *The solution* $w(t, t_0, w_0)$ *of dynamic initial value problem* (4.4) *exists for* $t \in \mathbb{T}$.

If F *is smooth enough to ensure the local existence, then* $[t_0, \infty)_{\mathbb{T}}$ *is the largest interval of existence of solution* $U(t, t_0, w_0)$ *of SDIVP* (4.3) *with* $\|U_0\| \leq w_0$.

Proof. Assume that $V(t) = V(t, t_0, U_0)$ is a solution of SDIVP (4.3) with $\|U_0\| \leq w_0$ existing on the largest interval $[t_0, T)_{\mathbb{T}}$. We will show that $T = \infty$. On the contrary, suppose there exists $\beta \leq T$ such that V is a solution of SDIVP (4.3) existing on interval $[t_0, \beta]_{\mathbb{T}}$ and β cannot be increased. Next, set

$$m(t) = D[V(t), \theta] \quad \text{for } t \in [t_0, \beta]_{\mathbb{T}}.$$

Employing the procedure used in the proof of Theorem 4.3, we obtain

$$m^{\Delta}(t) \leq g(t, m(t)), \quad t \in [t_0, \beta]_{\mathbb{T}}.$$

Since $\|U_0\| \leq w_0$, we have

$$D[V(t), \theta] \leq r(t, t_0, w_0), \quad t \in [t_0, \beta]_{\mathbb{T}}, \tag{4.6}$$

where $r(t, t_0, w_0)$ is the maximal solution of (4.3). Now, we select that $t_1, t_2 \in [t_0, \beta]_{\mathbb{T}}$ such that $t_1 < t_2$. Note that the solution of SDIVP (4.3) on $[t_1, t_2]_{\mathbb{T}}$ is given by

$$V(t) = V(t_1) + \int_{t_0}^{t} F(s, V(s)) \Delta_g s,$$

by means of the properties of Hausdorff metric, we have

$$D[V(t_1), V(t_2)] = D\left[V(t_1), V(t_1) + \int_{t_1}^{t_2} F(s, U) \Delta_g s\right]$$

$$= D\left[\theta, \int_{t_1}^{t_2} F(s, U) \Delta_g s\right]$$

$$\leq \int_{t_1}^{t_2} D[\theta, F(s, U(s))]\Delta s$$

$$\leq \int_{t_1}^{t_2} G(s, D[U(s), \theta])\Delta s.$$

In view of the nondecreasing property of G in w and from (4.6), it follows that

$$D[V(t_1), V(t_2)] \leq \int_{t_1}^{t_2} G(s, r(s, t_0, w_0))\Delta s$$

$$= r(t_2, t_0, w_0) - r(t_1, t_0, w_0).$$

If β is left-dense, then we allow $t_1, t_2 \to \beta$ in the above relation. Since

$$\lim_{t \to \beta^-} r(t)$$

exists and is finite, from the above inequality, in view of the Cauchy criterion of convergence, it follows that the limit of

$$\lim_{t \to \beta^-} V(t)$$

exists and is finite. We can define

$$V(\beta) = \lim_{t \to \beta^-} V(t).$$

Also, we observe that a solution can have finite escape time only before left-dense points $t \in \mathbb{T}$, since their neighborhoods contain infinitely many points to the left of t. Hence, it is sufficient that we are only allowed to suppose that β is a left-dense point of $\mathbb{T}$. Consequently, we consider $W_0 = V(\beta)$ as a new initial function at $t = \beta$. Then, by the assumption of local existence, there exists a solution $V(t, \beta, W_0)$ of SDIVP (4.3) on the interval $[\beta, \gamma]$ with $\gamma > \beta$. This implies that the solution V can be continued beyond β, which contradicts our assumption that β cannot be increased. This completes the proof. □

We present the following comparison theorem in terms of Lyapunov-like functions on time scales and this is very essential to investigate the stability criteria of SDE (4.3). To this end, we supplement some notions and definitions.

Atici *et al.* (2000) defined the Dini derivatives of a Lyapunov-like function on time scales $v \in C_{\mathrm{rd}}(\mathbb{T} \times \mathbb{R}^n, \mathbb{R}^+)$ along the solutions of SDIVP (4.3) when we restrict ourselves into single-valued mappings $U = u$ and $F = f$ by

$$D_{\Delta}^- v(t, u) = \lim_{\mu(t) \to 0} \inf \frac{v(t, u) - v(t - \mu(t), u - \mu(t) f(t, u))}{\mu(t)},$$

$$D_{\Delta}^+ v(t, u) = \lim_{\mu(t) \to 0} \sup \frac{v(t + \mu(t), u + \mu(t) f(t, u)) - v(t, u)}{\mu(t)}.$$

Now, to avoid the nonexistence of the above Dini derivatives when $\mu(t) \geq h$ (a positive constant), we present a class of new generalized Dini derivatives of the Lyapunov-like functions on $\mathbb{T}$. The following definition is due to Hong (2010).

Definition 4.2. For $A \in C_{\mathrm{rd}}(\mathbb{T}, K_c^1)$, $t \in \mathbb{T}$ and $\bar{v} \in C_{\mathrm{rd}}(\mathbb{T} \times K_c^1, \mathbb{R}^+)$, we call $\Delta^r \bar{v}(t, A)$ and $\Delta_r \bar{v}(t, A)$ the upper right (ur) and the lower right (lr) delta Dini derivatives of the function $\bar{v}$ at $(t, A(t))$, respectively, provided

$\Delta^r \bar{v}(t, A(t))$

$$= \begin{cases} \dfrac{\bar{v}(\sigma(t), A(\sigma(t))) - \bar{v}(t, A(t))}{\mu(t)} & \text{for } \sigma(t) > t \\[2ex] \lim_{s \to t^+} \sup \dfrac{\bar{v}(s, A(t) + (s - t)F(t, A(t))) - \bar{v}(t, A(t))}{s - t} & \text{for } \sigma(t) = t, \end{cases}$$

and

$\Delta_r \bar{v}(t, A(t))$

$$= \begin{cases} \dfrac{\bar{v}(\sigma(t), A(\sigma(t))) - \bar{v}(t, A(t))}{\mu(t)} & \text{for } \sigma(t) > t \\[2ex] \lim_{s \to t^+} \inf \dfrac{\bar{v}(s, A(t) + (s - t)F(t, A(t))) - \bar{v}(t, A(t))}{s - t} & \text{for } \sigma(t) = t. \end{cases}$$

Similarly, we call $\Delta^l \bar{v}(t, A(t))$ and $\Delta_l \bar{v}(t, A(t))$ the upper left (ul) and lower left (ll) delta derivatives of the function $\bar{v}$ at $(t, A(t))$, respectively, provided

$\Delta^l \bar{v}(t, A(t))$

$$= \begin{cases} \dfrac{\bar{v}(t, A(t)) - \bar{v}(\rho(t), A(\rho(t)))}{t - \rho(t)} & \text{for } t > \rho(t) \\[2ex] \lim_{s \to t^-} \sup \dfrac{\bar{v}(s, A(t) + (s - t)F(t, A(t))) - \bar{v}(t, A(t))}{s - t} & \text{for } \rho(t) = t, \end{cases}$$

and

$$\Delta_l \bar{v}(t, A(t))$$

$$= \begin{cases} \dfrac{\bar{v}(t, A(t)) - \bar{v}(\rho(t), A(\rho(t)))}{\nu(t)} & \text{for } t > \rho(t) \\[2ex] \displaystyle\liminf_{s \to t^-} \dfrac{\bar{v}(s, A(t) + (s-t)F(t, A(t))) - \bar{v}(t, A(t))}{s - t} & \text{for } \rho(t) = t. \end{cases}$$

Theorem 4.5. *Assume that $\bar{v}$ given as in Definition 4.2 satisfies*

$$\Delta^r \bar{v}(t, U(t)) \le G(t, \|U(t)\|) \quad \text{for } t \in \mathbb{T},$$

and

$$|\bar{v}(t, U(t)) - \bar{v}(t, V(t))| \le LD[U(t), V(t)] \quad \text{for } L \ge 0, \ t \in \mathbb{T},$$

where $G \in \mathcal{C}_{\mathrm{rd}}(\mathbb{T} \times \mathbb{R}^+, \mathbb{R})$ and $U(t) = U(t, t_0, U_0)$ and $V(t) = V(t, t_0, V_0)$ are solutions of SDIVP (4.3). Suppose $U(t)$ is such that $\bar{v}(t_0, U_0) \le w_0$ on $[t_0, \infty)_\mathbb{T}$. Then, we have

$$\bar{v}(t, U(t)) \le r(t, t_0, w_0) \quad \text{for } t \in [t_0, \infty)_\mathbb{T}, \tag{4.7}$$

where $r(t, t_0, w_0)$ is the maximal solution of DIVP (4.4) existing on $[t_0, \infty)_\mathbb{T}$.

Proof. Define $m(t) = \bar{v}(t, U(t))$, where $U(t) = U(t, t_0, U_0)$ is any solution of SDIVP (4.3) existing on $[t_0, \infty)_\mathbb{T}$. Then,

$$m(t_0) = \bar{v}(t_0, U_0) \le w_0,$$

Now, for $s \in \mathbb{T}$ with $s > t$, by our assumptions, it follows that

$$\begin{aligned} m(s) - m(t) &= \bar{v}(s, U(s)) - \bar{v}(t, U(t)) \\ &\le \bar{v}(s, U(s)) - \bar{v}(s, U(t) + (s-t)F(t, U(t))) \\ &\quad + \bar{v}(s, U(t) + (s-t)F(t, U(t))) - \bar{v}(t, U(t)) \\ &\le LD[U(s), U(t) + (s-t)F(t, U(t))] \\ &\quad + \bar{v}(s, U(t) + (s-t)F(t, U(t))) - \bar{v}(t, U(t)). \quad (4.8) \end{aligned}$$

Since $U(\cdot)$ is solution of SDIVP (4.3), the g-difference $Z(t) = U(s) -_g U(t)$ exists for $s, t \in \mathbb{T}$ and small $s - t > 0$. Hence, employing the properties of Hausdorff metric, we have

$$D[U(s), U(t) + (s - t)F(t, U(t))]$$
$$= D[U(t) + Z(t), U(t) + (s - t)F(t, U(t))]$$
$$= D[Z(t), (s - t)F(t, U(t))]$$
$$= D[U(s) -_g U(t), (s - t)F(t, U(t))].$$

This shows that

$$\frac{1}{s-t} D[U(s), U(t) + (s-t)F(t, U(t))] = D\left[\frac{U(s) -_g U(t)}{s - t}, F(t, U(t))\right].$$

Lending this to (4.8), we obtain

$$\frac{m(s) - m(t)}{s - t} \leq L D\left[\frac{U(s) -_g U(t)}{s - t}, F(t, U(t))\right]$$
$$+ \frac{\bar{v}(s, (s - t)F(t, U(t))) - \bar{v}(t, U(t))}{s - t}.$$

Employing the procedure used in the proof of Theorem 4.3, we obtain

$$\lim_{s \to t^+} \sup \frac{1}{s - t} D[U(s), U(t) + (s - t)F(t, U(t))]$$
$$= \lim_{s \to t^+} \sup D\left[\frac{U(s) -_g U(t)}{s - t}, F(t, U(t))\right]$$
$$= D[\Delta_g U(t), F(t, U(t))]$$
$$= 0.$$

Again, employing the procedure used in the proof of Theorem 4.3, and keeping in mind Definition 4.2 and Proposition 2.3(IV), that

$$m_+^\Delta(t) = \Delta^r(t, U(t))$$
$$\leq G(t, m(t)),$$

that is,

$$m_+^\Delta(t) \leq G(t, m(t)) \quad \text{for } t \in [t_0, \infty)_\mathbb{T}.$$

Consequently, by Theorem A.7 we arrive at the estimate

$$m(t) \leq r(t, t_0, w_0) \quad \text{for } t \in [t_0, \infty)_\mathbb{T},$$

where $r(t, t_0, w_0)$ is the maximal solution of DIVP (4.4) on $[t_0, \infty)_\mathbb{T}$. This completes the proof. $\qquad \square$

4.3.2   *Stability criteria*

We are now in a position to formulate the stability criteria for the solution of SDIVP (4.3). Throughout this section, we assume that $F(t, \theta) = \theta$ and the solution exists and is unique for all $t \in \mathbb{T}$ with $t \geq t_0$. Let us first define the stability of the trivial solution and fix our notations.

Definition 4.3. Let $U(t) = U(t, t_0, U_0)$. Then the trivial solution $U(t) \equiv \theta$ is said to be

   (I) equi-stable provided for each $\varepsilon > 0$ and $t_0 \in \mathbb{T}$, there exists $\delta = \delta(t_0, \varepsilon) > 0$ such that $\|U_0\| < \delta$ implies that

$$\|U(t)\| < \varepsilon \quad \text{for } t \in [t_0, \infty)_\mathbb{T}. \tag{4.9}$$

  (II) uniformly stable provided the $\delta = \delta(\varepsilon) > 0$, in (I), is independent of t_0.

 (III) equi-asymptotically stable provided (I) holds, and for any $\varepsilon > 0$, there exists $T > 0$ such that (4.9) holds for all $t \in [t_0 + T, \infty)_\mathbb{T}$.

 (IV) uniformly asymptotically stable provided (II) and (III) hold simultaneously.

Let

$$S(b) = \left\{ U \in K_c^1 \colon \|U\| \leq b \right\} \quad \text{for } b > 0,$$

and

$$\mathscr{K} = \left\{ \varphi \in C[[0, b), \mathbb{R}^+] \colon \varphi \text{ is increasing and } \varphi(0) = 0 \right\}.$$

For the sake of convenient, the following basic notions are also needed.

Definition 4.4. The function $\bar{v} \in \mathcal{C}_{\mathrm{rd}}(\mathbb{T} \times K_c^1, \mathbb{R})$ is said to be positive definite provided

 (i) $\bar{v}(t, \theta) = 0$ for all $t \in \mathbb{T}$; and

 (ii) there exists $\varphi \in \mathscr{K}$ such that $\bar{v}(t, A) \geq \varphi(\|A\|)$ for each $(t, A) \in \mathbb{T} \times S(b)$.

The function $\bar{v}$ is said to be negative definite provided $-\bar{v}$ is positive definite.

We begin with the following result that follows the corresponding result, i.e., Theorem A.8 with appropriate modifications.

Theorem 4.6. *Suppose that $F \in C_{rd}(\mathbb{T} \times K_c^1, K_c^1)$ and $G \in C_{rd}(\mathbb{T} \times \mathbb{R}^+, \mathbb{R})$ such that $g(t, 0) \equiv 0$ for all $t \in \mathbb{T}$. Moreover,*

(i) *there exists $\bar{v}$ given as in Definition 4.2, such that $\bar{v}(t, \theta) = 0$ for all $t \in \mathbb{T}$ and*

$$\Delta^r \bar{v}(t, U(t)) \leq G(t, \|U(t)\|) \quad \text{for } t \in \mathbb{T},$$

$$|\bar{v}(t, U(t)) - \bar{v}(t, W(t))| \leq LD[U(t), W(t)] \quad \text{for } L \geq 0, \ t \in \mathbb{T},$$

with $U(t) = U(t, t_0, U_0)$ and $W(t) = W(t, t_0, W_0)$.

(ii) *$G(t, w)$ is nondecreasing in w for each $t \in \mathbb{T}$.*

Then the equi-stability of the trivial solution of SDIVP

$$w^{\Delta}(t) = G(t, w), \quad w(t_0) = w_0 \geq 0, \quad t \in \mathbb{T} \tag{4.10}$$

implies the corresponding stability properties of the trivial solution of SDIVP (4.3).

Proof. Since $\bar{v}(t_0, \theta) = 0$ and $\bar{v}(t_0, A)$ is continuous with respect to A, for any $\varepsilon > 0$, there exists $\delta_1 = \delta_1(t_0, \varepsilon) > 0$ such that $\|U_0\| < \delta_1$ implies that $\bar{v}(t_0, U_0) \leq w_0$.

Let the trivial solution of (4.10) be equi-stable. Then, given $\varepsilon > 0$ and $t_0 \geq 0$, there exists $\delta_2 = \delta_2(t_0, \varepsilon) > 0$ such that

$$0 \leq w_0 < \delta_2 \quad \text{implies} \quad w(t) < \varphi(\varepsilon) \quad \text{for } t \in \mathbb{T}, \tag{4.11}$$

where $w(t) = w(t, t_0, w_0)$ is any solution of (4.10). We claim that, with these ε and $\delta = \min\{\delta_1, \delta_2\}$, the trivial solution of SDIVP (4.3) is also equi-stable. Suppose that this were false. Then there would exist $U(t) = U(t, t_0, U_0)$, a solution of SDIVP (4.3) with $\|U_0\| < \delta$, and $t_1 \in \mathbb{T}$, $t_1 > t_0$, such that

$$\varphi(\varepsilon) \leq \|U(t_1)\| \quad \text{and} \quad \|U(t)\| < \varphi(\varepsilon) \quad \text{for } t \in [t_0, t_1)_{\mathbb{T}}, \text{ and for some}$$
$$\varphi \in \mathcal{K}.$$

On the other hand, using (4.7) at $t = t_1$, we arrive at

$$\varphi(\varepsilon) \leq \|U(t_1)\|$$
$$\leq r(t_1, t_0, w_0)$$
$$< \varphi(\varepsilon),$$

which is contradiction. Hence the trivial solution of SDIVP (4.3) is also equi-stable. This completes the proof. $\square$

Theorem 4.7. *Suppose that $F \in \mathcal{C}_{\mathrm{rd}}(\mathbb{T} \times K_c^1, K_c^1)$. Moreover, there exists $\bar{v}$ given as in Definition 4.2 such that $\bar{v}$ is positive definite and*

$$\Delta^r \bar{v}(t, U(t)) \leq 0 \quad \text{for } t \in \mathbb{T}, \quad \text{and}$$

$$|\bar{v}(t, U(t)) - \bar{v}(t, W(t))| \leq LD[U(t), W(t)] \quad \text{for } L \geq 0, \ t \in \mathbb{T},$$

where $U(t) = U(t, t_0, U_0)$ and $W(t) = W(t, t_0, U_0)$. Then the trivial solution of SDIVP (4.3) is equi-stable.

Proof. Define $m(t) = \bar{v}(t, U(t))$ for $t \in \mathbb{T}$. Then, keeping in mind the fact that $\bar{v}$ is positive definite, we have

$$m(t) = \bar{v}(t, U(t)) \geq \varphi(\|U(t)\|)$$

for some $\varphi \in \mathcal{K}$. Since $\bar{v}(t_0, \theta) = 0$ and $\bar{v}(t_0, A)$ is continuous with respect to A, for any $\varepsilon > 0$, there exists $\delta = \delta(t_0, \varepsilon) > 0$ ($\delta \leq \varphi(\varepsilon)$) such that $\|U_0\| < \delta$ implies that

$$\bar{v}(t_0, U_0) < \varphi(\varepsilon),$$

that is,

$$m(t_0) = \bar{v}(t_0, U_0) < \varphi(\varepsilon).$$

On the other hand, DIVP

$$u^\Delta = 0, \quad u(t_0) = \bar{v}(t_0, U_0),$$

has only solution $u \equiv \bar{v}(t_0, U_0)$. Now, employing the procedure used in the proof of Theorem 4.5, the hypothesis $\Delta^r \bar{v}(t, A) \leq 0$ guarantees

$$m(t) \leq \bar{v}(t_0, U_0), \quad t \in [t_0, \infty)_{\mathbb{T}}. \tag{4.12}$$

Therefore, for all $t \in [t_0, \infty)_{\mathbb{T}}$, we have

$$\varphi(\|U(t)\|) \leq \bar{v}(t, U(t))$$

$$\leq \bar{v}(t_0, U_0)$$

$$< \varphi(\varepsilon).$$

Since φ is increasing, we infer U satisfies (4.9), i.e., the trivial solution of SDIVP (4.3) is equi-stable as desired. This completes the proof. $\square$

We will next consider the uniform stability criteria.

Theorem 4.8. *Suppose that $F \in \mathcal{C}_{\mathrm{rd}}(\mathbb{T} \times K_c^1, K_c^1)$. Moreover, there exists $\bar{v}$ given as in Definition 4.2 such that $\bar{v}$ is positive definite and*

$$\Delta^r \bar{v}(t, U(t)) \leq 0 \quad \text{for } t \in \mathbb{T},$$

and

$$|\bar{v}(t, U(t)) - \bar{v}(t, W(t))| \leq L D[U(t), W(t)] \quad \text{for } L \geq 0,\ t \in \mathbb{T},$$

where $U(t) = U(t, t_0, U_0)$ and $W(t) = W(t, t_0, U_0)$. Also, there exists $\psi \in \mathscr{K}$ such that

$$\bar{v}(t, A) \leq \psi(\|A\|),$$

where $(t, A) \in \mathbb{T} \times S(b)$. Then the trivial solution of SDIVP (4.3) is uniformly stable.

Proof. Theorem 4.7 guarantees that the trivial solution of SDIVP (4.3) is stable. We next prove the uniform stability criteria. By our assumptions, there exists $\varphi \in \mathscr{K}$ such that

$$\varphi(\|A\|) \leq \bar{v}(t, A)$$
$$\leq \psi(\|A\|)$$

for all $(t, A) \in \mathbb{T} \times S(b)$. Now, for any $\varepsilon > 0$ satisfying

$$\psi^{-1}(\varphi(\varepsilon)) < b,$$

set $\delta = \psi^{-1}(\varphi(\varepsilon))$. Since $\varphi, \psi \in \mathscr{K}$, $\delta = \delta(\varepsilon)$ exists uniquely and is independent of t_0. Thus, keeping in mind $\|U_0\| < \delta$, from (4.12), it follows that, for $t \in [t_0, \infty)_{\mathbb{T}}$,

$$\varphi(\|U(t)\|) \leq \bar{v}(t, U(t))$$
$$\leq \bar{v}(t_0, U_0)$$
$$\leq \psi(\|U_0\|)$$
$$< \psi(\delta).$$

This yields that

$$\|U(t)\| < \varphi^{-1}(\psi(\delta)) = \varepsilon \quad \text{for } t \in [t_0, \infty)_{\mathbb{T}}.$$

Hence, the trivial solution of SDIVP (4.3) is uniformly stable. This completes the proof. □

The next result provides sufficient condition for equi-asymptotic stability criteria.

Theorem 4.9. *Let the assumptions of Theorem 4.7 hold except that the estimate $\Delta^r \bar{v}(t, A) \leq 0$ be strengthened to*

$$\Delta^r \bar{v}(t, A) \leq -\phi(\|A\|) \tag{4.13}$$

for $(t, A) \in \mathbb{T} \times K_c(\mathbb{R})$, where $\phi \in \mathcal{K}$ is given. Then, the trivial solution of SDIVP (4.3) is equi-asymptotically stable.

Proof. Clearly, Theorem 4.7 guarantees that the trivial solution of SDIVP (4.3) is equi-stable. Thus, given any $\varepsilon > 0$, there is $\delta = \delta(t_0, \varepsilon) > 0$ such that $\|U_0\| < \delta$ implies that

$$\|U(t)\| < \varepsilon \quad \text{for } t \in [t_0, \infty)_{\mathbb{T}}.$$

From our assumptions, it follows that

$$\bar{v}(t, U(t)) + \int_{t_0}^t \phi(\|U(s)\|)\Delta_g s \leq \bar{v}(t_0, U_0) \quad \text{for } t \in [t_0, \infty)_{\mathbb{T}}.$$

Note that the inequalities $\bar{v}(t, U(t)) \geq 0$ and $\Delta^r \bar{v}(t, A) < 0$ guarantee that the function $\bar{v}_U(t) = \bar{v}(t, U(t))$ is decreasing in $t \in \mathbb{T}$. This shows that there is $\alpha \in \mathbb{R}$ such that

$$\lim_{t \to \infty} \bar{v}_U(t) = \alpha.$$

We prove that $\alpha = 0$. On the contrary, suppose $\alpha > 0$. Then, by means of the fact that $\bar{v}_U$ is decreasing, we have

$$\bar{v}_U(t) \geq \alpha > 0 \quad \text{for all } t \in [t_0, \infty)_{\mathbb{T}}.$$

Also, from the continuity of $\bar{v}_U(t) = \bar{v}(t, U(t))$ with respect to U and $\bar{v}(t, 0) = 0$, there is a positive constant ξ such that $\|U(t)\| > \xi$ for each $t \in [t_0, \infty)_{\mathbb{T}}$. This, together with $\psi \in \mathcal{K}$, yields that

$$\bar{v}(t, U(t)) \leq \bar{v}(t_0, U_0) - \int_{t_0}^t \phi(\|U(s)\|)\Delta s$$

$$\leq \bar{v}(t_0, U_0) - \phi(\xi)(t - t_0)$$

for all $t \in [t_0, \infty)_{\mathbb{T}}$. Now, taking $t \in \mathbb{T}$ to be large enough, we obtain that $\bar{v}_U(t) < 0$, a contradiction. Hence, $\alpha = 0$, that is,

$$\lim_{t \to \infty} \bar{v}_U(t) = 0.$$

Next, we prove that

$$\lim_{t \to \infty} \|U(t)\| = 0.$$

If this were false, there would exists $\varepsilon_0 > 0$ such that $\|U(t_k)\| > \varepsilon_0$ for some $t_k \in \mathbb{T}$ with $t_k \geq k$, $k \in \mathbb{N}$. From this, combining the fact that $\bar{v}$ is positive definite, there would exists $\varphi \in \mathscr{K}$ such that

$$\bar{v}(t_k, U(t_k)) \geq \varphi(\|U(t_k)\|)$$
$$\geq \varphi(\varepsilon_0)$$
$$> 0$$

for $k \in \mathbb{N}$. This contradicts the fact

$$\lim_{t \to \infty} \bar{v}(t, U(t)) = 0.$$

Hence,

$$\lim_{t \to \infty} \|U(t)\| = 0$$

which guarantees that the trivial solution of SDIVP (4.3) is equi-asymptotically stable. This completes the proof. $\qquad\square$

The next theorem presents the sufficient conditions for uniformly asymptotic stability.

Theorem 4.10. *Suppose the assumptions of Theorem 4.8 hold except that the estimate $\Delta^r \bar{v}(t, A) \leq 0$ be strengthened to (4.13). Then the trivial solution of SDIVP (4.3) is uniformly asymptotically stable.*

Proof. Theorem 4.8 guarantees that the trivial solution of SDIVP (4.3) uniformly stable. By our assumptions, there exist functions $\phi, \varphi, \psi \in \mathscr{K}$ such that

$$\varphi(\|A\|) \leq \bar{v}(t, A) \leq \psi(\|A\|) \quad \text{for all } (t, A) \in \mathbb{T} \times S(b), \qquad (4.14)$$

and

$$\Delta^r \bar{v}(t, U(t)) \leq -\phi(\|U(t)\|) \leq -\phi(\psi^{-1}(\bar{v}(t, U(t)))) < 0. \qquad (4.15)$$

Rewriting (4.15), we get

$$\frac{\Delta^r \bar{v}_U(t)}{\phi(\psi^{-1}(\bar{v}_U(t)))} \leq -1.$$

This yields that

$$\int_{\bar{v}_U(t_0)}^{\bar{v}_U(t)} \frac{1}{\phi(\psi^{-1}(\bar{v}_U(s)))} \Delta \bar{v}_U(s) \leq -(t - t_0),$$

i.e.,

$$\int_{\bar{v}_U(t)}^{\bar{v}_U(t_0)} \frac{1}{\phi(\psi^{-1}(\bar{v}_U(s)))} \Delta \bar{v}_U(s) \geq (t - t_0).$$

Also, from (4.14), it follows that

$$\bar{v}_U(t_0) \leq \psi(\|U_0\|) \leq \psi(b).$$

Again, for any $\varepsilon > 0$ $(\varepsilon < b)$, from (4.14), we have

$$\int_{\varphi(\|U(t)\|)}^{\psi(b)} \frac{1}{\phi(\psi^{-1}(\bar{v}_U(s)))} \Delta \bar{v}_U(s) = \int_{\varphi(\|U(t)\|)}^{\psi(\varepsilon)} \frac{1}{\phi(\psi^{-1}(\bar{v}_U(s)))} \Delta \bar{v}_U(s)$$

$$+ \int_{\varphi(\varepsilon)}^{\psi(b)} \frac{1}{\phi(\psi^{-1}(\bar{v}_U(s)))} \Delta \bar{v}_U(s)$$

$$\geq \int_{\bar{v}_U(t)}^{\bar{v}_U(t_0)} \frac{1}{\phi(\psi^{-1}(\bar{v}_U(s)))} \Delta \bar{v}_U(s)$$

$$\geq (t - t_0).$$

Take

$$T = T(\varepsilon, b) > \int_{\varphi(\varepsilon)}^{\psi(b)} \frac{1}{\phi(\psi^{-1}(\bar{v}_U(s)))} \Delta \bar{v}_U(s).$$

Then, this T is independent of t_0 and U_0, and

$$\int_{\varphi(\|U(t)\|)}^{\psi(b)} \frac{1}{\phi(\psi^{-1}(\bar{v}_U(s)))} \Delta \bar{v}_U(s)$$

$$\geq t - t_0 - \int_{\varphi(\varepsilon)}^{\psi(b)} \frac{1}{\phi(\psi^{-1}(\bar{v}_U(s)))} \Delta \bar{v}_U(s)$$

$$> t - t_0 - T.$$

That is,

$$\int_{\varphi(\|U(t)\|)}^{\psi(b)} \frac{1}{\phi(\psi^{-1}(\bar{v}_U(s)))} \Delta \bar{v}_U(s) \geq 0$$

for all $t \in (t_0 + T, \infty)_{\mathbb{T}}$. This yields that

$$\varphi(\|U(t)\|) < \varphi(\varepsilon)$$

for all $t \in (t_0 + T, \infty)_{\mathbb{T}}$. In view of the monotonicity of φ, we obtain

$$\|U(t)\| < \varepsilon \quad \text{for all } t \in (t_0 + T, \infty)_{\mathbb{T}}.$$

This guarantees that the trivial solution of SDIVP (4.3) is uniformly asymptotic stable. This completes the proof. $\square$

Finally, we consider the unstable criteria of the trivial solution of SDIVP (4.3).

Theorem 4.11. *Suppose that $\bar{v}$ given as in Definition 4.2 such that $\bar{v}(t, \theta) = 0$ and for given $c > 0$, there exists $A \in S(c)$ such that $\bar{v}(t, A) > 0$, where $t \in \mathbb{T}$. Moreover, if $\Delta^r \bar{v}(t, A)$ is positive definite, and, there is $\psi \in \mathcal{K}$ such that*

$$\bar{v}(t, A) \leq \psi(\|A\|) \quad \text{for all } (t, A) \in \mathbb{T} \times S(b),$$

then the trivial solution of SDIVP (4.3) is unstable.

Proof. For given $\delta > 0$, by our assumptions, there is $W_0 \in S(\delta)$ (i.e., $\|W_0\| < \delta$) such that $\bar{v}(t_0, W_0) > 0$. If the trivial solution of SDIVP (4.3) is stable, then for given $\varepsilon > 0$ ($\varepsilon < b$), there is $\delta > 0$ such that $\|W_0\| < \delta$ implies

$$\|U(t)\| < \varepsilon \quad \text{for all } t \in [t_0, \infty)_{\mathbb{T}},$$

where $U(t) = U(t, t_0, W_0)$. Since $\Delta^r \bar{v}(t, U)$ is positive definite, $\bar{v}(t, U(t))$ is increasing. We have

$$\bar{v}(t, U(t)) \geq \bar{v}(t_0, W_0) > 0 \quad \text{for all } t > t_0.$$

This yields that

$$\psi(\|U(t)\|) \geq \bar{v}(t, U(t)) \geq \bar{v}(t_0, W_0) > 0 \quad \text{for all } t > t_0,$$

subsequently,

$$\|U(t)\| \geq \psi^{-1}(\bar{v}(t_0, W_0)) := \alpha > 0.$$

Again, applying the hypothesis that $\Delta^r \bar{v}(t, U)$ is positive definite, there exists $\varphi \in \mathcal{K}$ such that

$$\Delta^r \bar{v}(t, U(t)) \geq \varphi(\|U(t)\|)$$

for $(t, U(t)) \in \mathbb{T} \times K_c^1$. Now, integrating this inequality with $t > t_0$ yields

$$\bar{v}(t, U(t)) \geq \bar{v}(t_0, W_0) + \int_{t_0}^t \varphi(\|U(s)\|)\Delta s$$

$$\geq \bar{v}(t_0, W_0) + \varphi(\alpha)(t - t_0).$$

Since $\|U(t)\| < \varepsilon$, we have

$$\psi(\varepsilon) \geq \bar{v}(t_0, W_0) + \varphi(\alpha)(t - t_0).$$

Since $\psi \in \mathcal{K}$ and $\varepsilon > 0$ is arbitrarily small, the above inequality yields a contradiction for t approaching infinity. Hence, the trivial solution of SDIVP (4.3) is unstable. This completes the proof. $\square$

Example 4.1. Consider the dynamic initial value problem

$$u^\Delta = \ominus u, \quad u(0) = u_0 \in \mathbb{R}, \quad 0 \in \mathbb{T},$$

and the corresponding set dynamic initial value problem

$$\Delta_g \bar{U} = \ominus U, \quad U(0) = U_0 \in K_c^1. \tag{4.16}$$

By means of Theorem 2.1 and the properties of Hausdorff metric, we see that the solution of (4.16) is unique whenever it exists. We next prove that the values of the solution of (4.16) are interval function. In fact, let $U = [u_1, u_2]$, where

$$u_1(t) = \frac{1}{2}[u_{10} + u_{20}]e_1(0, t) + \frac{1}{2}[u_{10} \ominus u_{20}]e_{\ominus 1}(t, 0)$$

and

$$u_2(t) = \frac{1}{2}[u_{10} + u_{20}]e_1(0, t) + \frac{1}{2}[u_{20} \ominus u_{10}]e_{\ominus 1}(t, 0).$$

In virtue of Theorem 1.6, we have

$$u_1^\Delta(t) = \ominus \frac{1}{2}[u_{10} + u_{20}]e_1(0,t) + \frac{1}{2}[u_{10} \ominus u_{20}]e_{\ominus 1}(t,0) = \ominus u_2(t)$$

and

$$u_2^\Delta(t) = \ominus \frac{1}{2}[u_{10} + u_{20}]e_1(0,t) + \frac{1}{2}[u_{20} \ominus u_{10}]e_{\ominus 1}(t,0) = \ominus u_1(t).$$

This shows that the vector function (u_1, u_2) is the solution of the system of dynamic equations

$$u_1^\Delta(t) = \ominus u_2(t), \quad u_1(0) = u_{10},$$
$$u_2^\Delta(t) = \ominus u_1(t), \quad u_2(0) = u_{20}.$$

Conclusively, $U = [u_1, u_2]$ is a solution of SDIVP (4.16) with the initial value $U_0 = [u_{10}, u_{20}]$.

Conclusion 1. Given $U_0 \in K_c^1$ and we choose

$$\bar{v}(t, A(t)) = \|A(t)\| \quad \text{for } t \in \mathbb{T}.$$

Then, for $U(t) = U(t, t_0, U_0)$, a solution of (4.16) corresponding the initial value $(0, U_0)$, we have

$$\Delta^r \bar{v}(t, U(t))$$

$$= \begin{cases} \dfrac{\|U(\sigma(t))\| - \|U(t)\|}{\mu(t)} & \text{for } \sigma(t) > t \\[3mm] \lim\limits_{s \to t^+} \sup \dfrac{\|U(t) + (s-t)(\ominus U(t))\| - \|U(t)\|}{s-t} \\[3mm] \quad = -\frac{1}{1+\mu(t)}\|U(t)\| = -\|U(t)\| & \text{for } \sigma(t) = t. \end{cases}$$

Now, take

$$G(t, w) = \frac{1}{1 + \mu(t)} w.$$

If $t \in \mathbb{T}$ is right-dense, then

$$\Delta^r \bar{v}(t, U(t)) = -\|U(t)\|$$
$$\leq G(t, \|U(t)\|).$$

If t is right-scattered, then

$$\Delta^r \bar{v}(t, U(t)) = \frac{\|U(\sigma(t))\| - \|U(t)\|}{\mu(t)}$$

$$\leq \frac{\|U(\sigma(t)) - U(t)\|}{\mu(t)}$$

$$= \|\Delta_g U(t)\|$$

$$= \|\ominus U(t)\|$$

$$= \frac{\|U(t)\|}{1 + \mu(t)}$$

$$= G(t, \|U(t)\|).$$

From Theorem 4.6, it follows that the equi-stability of trivial solution of (4.4) with

$$G(t, w) = \frac{1}{1 + \mu(t)} w$$

and $w_0 = \|U_0\|$ implies the corresponding equi-stability of the trivial solution of SDIVP (4.16).

Conclusion 2. On the basis of the above discussion, assuming $u_{10} \neq -u_{20}$, we obtain that

$$U(t) = \frac{1}{2} [u_{10} - u_{20}, u_{20} - u_{10}] e_{\ominus 1}(t, 0)$$

$$+ \frac{1}{2} [u_{10} + u_{20}, u_{20} + u_{10}] e_1(0, t).$$

By the analogical argument of set differential equations discussed in Lakshmikantham *et al.* (2004), for any general initial value U_0, the solution of SDIVP (4.16) contains both the desired and the undesired parts compared to the solution of the corresponding differential equation. Choose the appropriate initial value $U_0 = [u_{10}, u_{20}]$, say $U_0 = [c, c]$ for some real number c, such that the term with $e_{\ominus 1}(t, 0)$ is eliminated and only the desirable part of the solution compared with the dynamic equation is retained. In this case, note that $1 + \mu(t) > 0$, from Theorems 1.6 and 1.7, we have

$$e_1(0, t) > 0 \quad \text{and} \quad e_1^{\Delta}(0, \cdot) < 0.$$

Hence,

$$\Delta^r \bar{v}(t, U(t)) \leq 0 \quad \text{with } \bar{v}(t, A(t)) = \|A(t)\|.$$

Now, Theorem 4.7 guarantees that the trivial solution of SDIVP (4.16) is equi-stable.

Let $\psi \in \mathcal{K}$ with $\psi(w) = 2w$ for $w \in \mathbb{R}^+$. Then, Theorem 4.8 guarantees that the trivial solution of SDIVP (4.16) is uniformly stable. From the above discussion, it follows that

$$\|U(t)\| = |c|e_1(0, t) > 0$$

and $\|U(t)\|$ is decreasing. Now, we take

$$\phi(t) = \begin{cases} \dfrac{\|U(t)\| - \|U(\sigma(t))\|}{\mu(t)} & \text{for } \sigma(t) > t \\ \frac{1}{2}\|U(t)\| & \text{for } \sigma(t) = t. \end{cases}$$

Then, $\phi \in \mathcal{K}$ and the inequality (4.13) holds. Therefore, Theorem 4.10 guarantees that the trivial solution of SDIVP (4.16) is uniformly asymptotically stable.

Example 4.2. Consider the set dynamic initial value problem

$$\Delta_g U = \lambda(t)U, \quad U(0) = U_0, \tag{4.17}$$

which is generated by the dynamic initial value problem

$$u^\Delta = \lambda(t)u, \quad u(0) = u_0,$$

where $\lambda > 0$ is real-valued function defined on $\mathbb{T}^+ := \mathbb{R}^+ \cap \mathbb{T}$ such that $\lambda \in \mathcal{R}_1^+$. Then, with similar computation, we see that

$$U(t) = U_0 e_\lambda(t, 0) \quad \text{for } t \in \mathbb{T}^+,$$

is the unique solution of SDIVP (4.17) corresponding the initial value $(0, U_0)$ for any $U_0 \in K_c^1$. Since $e_\lambda(0, 0) = 1$ and

$$e_\lambda^\Delta(t, 0) = \lambda(t)e_\lambda(t, 0) > 0,$$

we obtain that $e_\lambda(t, 0)$ is increasing and $e_\lambda(t, 0) \geq 1$ on $\mathbb{T}^+$. Let $\bar{v}(t, A) = \|A\|$ for $A \in K_c^1$. Then,

$$\bar{v}(t, U(t)) = \|U_0\|e_\lambda(t, 0)$$

provided U is solution of SDIVP (4.17). It can be seen that the conditions of Theorem 4.11 are satisfied when $\|U_0\| > 0$. Consequently, the trivial solution of SDIVP (4.17) is unstable.

4.4 Exponential Stability and Generalization

In this section, we shall develop the exponential stability, exponentially asymptotic stability, uniform stability, and uniformly exponentially asymptotic stability, and their generalization for the trivial solution of SDIVP (4.3).

4.4.1 *Exponential stability*

As before, we assume that the time scale $\mathbb{T}$ has a minimal element $t_0 \geq 0$ and is not bounded above, and we adopt the Lyapunov functional approach. First, we define the Lyapunov-like function and discuss its properties.

Definition 4.5. A function $\bar{v} \colon \mathbb{T} \times K_c^1 \to \mathbb{R}^+$ is said to be Lyapunov-like function on $\mathbb{T} \times K_c^1$ provided that $\bar{v} \in \mathcal{C}_{\mathrm{rd}}^1(\mathbb{T} \times K_c^1, \mathbb{R}^+)$.

Theorem 4.12. *If $\bar{v}$ is the Lyapunov-like function, then its ur and lr delta derivatives exist for $(t, A) \in \mathbb{T} \times \mathcal{C}_{\mathrm{rd}}(\mathbb{T}, K_c^1)$. Moreover, for given A, we have*

$$\bar{v}^\Delta(t) = \Delta^r \bar{v}(t, A(t)) = \Delta_r \bar{v}(t, A(t)),$$

where $\bar{v}(t) = \bar{v}(t, A(t))$ for $t \in \mathbb{T}$.

Proof. If t is right-scattered, Definition 4.2 guarantees that

$$\Delta^r \bar{v}(t, A(t)) = \Delta_r \bar{v}(t, A(t))$$
$$= \bar{v}^\Delta(t, A(t))$$
$$= \frac{\bar{v}(\sigma(t), A(\sigma(t))) - \bar{v}(t, A(t))}{\mu(t)}.$$

If t is right-dense, i.e., $\sigma(t) = t$, then we obviously have

$$\liminf_{s \to t^+} \frac{\bar{v}(s, A(t) + (s-t)F(t, A(t))) - \bar{v}(t, A(t))}{s - t}$$
$$\leq \lim_{s \to t^+} \frac{\bar{v}(s, A(t) + (s-t)F(t, A(t))) - \bar{v}(t, A(t))}{s - t}$$
$$\leq \limsup_{s \to t^+} \frac{\bar{v}(s, A(t) + (s-t)F(t, A(t))) - \bar{v}(t, A(t))}{s - t}.$$

For the sake of convenience, we set

$$\phi(r) = \frac{\bar{v}(r, A(t) + (r - t)F(t, A(t))) - \bar{v}(t, A(t))}{r - t} \quad \text{for } t \in \mathbb{T},$$

and

$$w(u) = \sup_{r \in [t,u]_\mathbb{T}} \phi(r) \quad \text{and} \quad l(u) = \inf_{r \in [t,u]_\mathbb{T}} \phi(r) \quad \text{for } u \in [t, s]_\mathbb{T}.$$

Since t is right-dense, there is a convergent sequence $\{s_n\}_{n \in \mathbb{N}}$ in $\mathbb{T}$ with $s_n \geq s_{n+1} > t$ such that

$$\lim_{n \to \infty} s_n = t.$$

Let $s \in \mathbb{T}$ with $s > s_1$. Then, $[t, s_1]_\mathbb{T} \subseteq [t, s]_\mathbb{T}$ and there is $a_1 \in [t, s_1]_\mathbb{T}$ such that

$$\phi(a_1) = w(s_1).$$

Similarly, we can get that $[t, s_n]_\mathbb{T} \subseteq [t, s_{n-1}]_\mathbb{T}$ and there is $a_n \in [t, s_n]_\mathbb{T}$ such that

$$\phi(a_n) = w(s_n).$$

Therefore, we obtain another convergent sequence $\{a_n\}_{n \in \mathbb{N}}$ with

$$\lim_{n \to \infty} a_n = t.$$

From our assumptions, it follows that

$$[t, s_1]_\mathbb{T} \supseteq [t, s_2]_\mathbb{T} \supseteq \cdots \supseteq [t, s_n]_\mathbb{T} \supseteq \cdots .$$

Hence,

$$w(s_1) \geq w(s_2) \geq \cdots \geq w(s_n) \geq \cdots .$$

On the one hand, we have

$$\lim_{n \to \infty} w(s_n) = \lim_{n \to \infty} \phi(a_n)$$

$$= \lim_{n \to \infty} \frac{\bar{v}(a_n, A(t) + (a_n - t)F(t, A(t))) - \bar{v}(t, A(t))}{a_n - t}$$

$$= \bar{v}^\Delta(t).$$

On the other hand, the definition of ω implies that

$$\lim_{n \to \infty} \omega(s_n) = \lim_{n \to \infty} \sup_{r \in [t, s_n]_{\mathbb{T}}} \phi(r)$$

$$= \lim_{n \to \infty} \sup_{r \in [t, s_n]_{\mathbb{T}}} \frac{\bar{v}(r, A(t) + (r - t)F(t, A(t))) - \bar{v}(t, A(t))}{r - t}$$

$$= \Delta^r \bar{v}(t, A(t)).$$

Consequently,

$$\bar{v}^\Delta(t) = \Delta^r \bar{v}(t, A(t)) \quad \text{for } t \in \mathbb{T}.$$

We can similarly find $b_n \in [t, s_n]$ such that

$$\phi(b_n) = l(s_n) \quad \text{for } n \in \mathbb{N}. \tag{4.18}$$

Therefore, again we obtain convergent sequence $\{b_n\}_{n \in \mathbb{N}}$ such that

$$\lim_{n \to \infty} b_n = t.$$

On the one hand, we have

$$\lim_{n \to \infty} l(s_n) = \lim_{n \to \infty} \phi(b_n)$$

$$= \lim_{n \to \infty} \frac{\bar{v}(b_n, A(t) + (b_n - t)F(t, A(t))) - \bar{v}(t, A(t))}{b_n - t}$$

$$= \bar{v}^\Delta(t).$$

On the other hand, the definition of l implies that

$$\lim_{n \to \infty} l(s_n) = \lim_{n \to \infty} \inf_{r \in [t, s_n]_{\mathbb{T}}} \phi(r)$$

$$= \lim_{n \to \infty} \inf_{r \in [t, s_n]_{\mathbb{T}}} \frac{\bar{v}(r, A(t) + (r - t)F(t, A(t))) - \bar{v}(t, A(t))}{r - t}$$

$$= \Delta_r \bar{v}(t, A(t)).$$

This implies that

$$\bar{v}^\Delta(t) = \Delta_r \bar{v}(t, A(t)) \quad \text{for } t \in \mathbb{T}.$$

This completes the proof. $\square$

Definition 4.6. Let $U(t) = U(t, t_0, U_0)$, $t \in \mathbb{T}$. Then the trivial solution of SDIVP (4.3) is said to be

(I) exponentially stable on $\mathbb{T}$ provided there are positive constants d, M and a function $\varrho \colon \mathbb{R}^+ \times \mathbb{T} \to \mathbb{R}^+$ such that

$$\|U(t, t_0, U_0)\| \le \varrho(\|U_0\|, t_0)(e_{\ominus M}(t, t_0))^d \quad \text{for } t \in [t_0, \infty)_{\mathbb{T}}. \tag{4.19}$$

(II) uniformly exponentially stable provided the function ϱ, in (I), is independent of t_0.

(III) exponentially asymptotically stable provided (I) holds and also, for any given $\varepsilon > 0$, there is a positive real number T such that

$$\|U(t, t_0, U_0)\| < \varepsilon \quad \text{for all } t \in [t_0 + T, \infty)_{\mathbb{T}},$$

(IV) uniformly exponentially asymptotically stable provided both (II) and (III) hold simultaneously.

We are now in a position to present results of exponential stability.

Theorem 4.13. *Assume that* $\bar{v} \in \mathcal{C}^1_{\mathrm{rd}}(\mathbb{T} \times K^1_c, \mathbb{R}^+)$ *and* $U \in \mathcal{C}_{\mathrm{rd}}(\mathbb{T}, K^1_c)$ *satisfy the following conditions.*

(i) *There exist strictly increasing continuous functions* $\omega, \phi \colon [0, \infty) \to [0, \infty)$ *such that*

$$\omega(\|U(t)\|) \le \bar{v}(t, U(t)) \le \phi(\|U(t)\|) \quad \text{for all } t \in \mathbb{T}.$$

(ii) *There exist a nonincreasing continuous function* $\psi \colon [0, \infty) \to [-\infty, 0]$ *and nonnegative constants* L, M, *and* α *with* $\alpha \ge M$ *such that*

$$\bar{v}^\Delta(t, U(t)) \le \frac{\psi(\|U(t)\|) - L(M \ominus \alpha)(t)e_{\ominus \alpha}(t, 0)}{1 + M\mu(t)}$$

and

$$\psi(\phi^{-1}(\bar{v}(t, U(t)))) + M\bar{v}(t, U(t)) \le 0,$$

where ϕ *is given as in* (i).

Then all solutions of SDIVP (4.3) *satisfy*

$$\|U(t)\| \le \omega^{-1}((\bar{v}(t_0, U_0) + L)e_{\ominus M}(t, t_0)) \quad \text{for all } t \in \mathbb{T}.$$

Proof. By means of our assumptions and Theorem 4.12, we have

$$[\bar{v}(t, U(t))e_M(t, 0)]^{\Delta}$$
$$= \bar{v}^{\Delta}(t, U(t))e_M(\sigma(t), 0) + M\bar{v}(t, U(t))e_M(t, 0)$$
$$\leq (\psi(\|U(t)\|) - L(M \ominus \alpha)(t)e_{\ominus\alpha}(t, 0))e_M(t, 0)$$
$$+ M\bar{v}(t, U(t))e_M(t, 0)$$
$$\leq (\psi(\phi^{-1}(\bar{v}(t, U(t)))) + M\bar{v}(t, U(t))$$
$$- L(M \ominus \alpha)(t)e_{\ominus\alpha}(t, 0))e_M(t, 0)$$
$$\leq -L(M \ominus \alpha)(t)e_{\ominus\alpha}(t, 0)e_M(t, 0)$$
$$= -L(M \ominus \alpha)(t)e_{M\ominus\alpha}(t, 0).$$

That is,

$$[\bar{v}(t, U(t))e_M(t, 0)]^{\Delta} \leq -L(M \ominus \alpha)(t)e_{M\ominus\alpha}(t, 0) \quad \text{for } t \in \mathbb{T}.$$

Integrating this inequality from t_0 to t and observing $U_0 = U(t_0)$, we obtain

$$\bar{v}(t, U(t))e_M(t, 0) \leq \bar{v}(t_0, U_0)e_M(t_0, 0) - Le_{M\ominus\alpha}(t, 0) + Le_{M\ominus\alpha}(t_0, 0)$$
$$\leq \bar{v}(t_0, U_0)e_M(t_0, 0) + Le_{M\ominus\alpha}(t_0, 0)$$
$$\leq (\bar{v}(t_0, U_0) + L)e_M(t_0, 0).$$

This implies that

$$\bar{v}(t, U(t)) \leq (\bar{v}(t_0, U_0) + L)e_M(t_0, 0)e_{\ominus M}(t, 0)$$
$$= (\bar{v}(t_0, U_0) + L)e_{\ominus M}(t, t_0).$$

From this and the condition (i), it follows that

$$\|U(t)\| \leq \omega^{-1}((\bar{v}(t_0, U_0) + L)e_{\ominus M}(t, t_0)) \quad \text{for all } t \in \mathbb{T}.$$

This completes the proof.   □

Theorem 4.14. *Assume that* $\bar{v} \in C_{\mathrm{rd}}^1(\mathbb{T} \times K_c^1, \mathbb{R}^+)$ *and* $U \in C_{\mathrm{rd}}(\mathbb{T}, K_c^1)$ *satisfy the following conditions.*

(i) *There exist positive functions* λ_1, λ_2 *and positive constants* p, q *such that*

$$\lambda_1(t)(\|U(t)\|)^p \leq \bar{v}(t, U(t)) \leq \lambda_2(t)(\|U(t)\|)^q \quad \text{for all } t \in \mathbb{T}.$$

(ii) *There exist a positive function λ_3, nonnegative constant L, and positive constants r and α with*

$$\alpha > M := \inf_{t \geq 0} \frac{\lambda_3(t)}{[\lambda_2(t)]^{r/q}} > 0$$

such that

$$\bar{v}^{\Delta}(t, U(t)) \leq \frac{-\lambda_3(t)(\|U(t)\|)^r - L(M \ominus \alpha)(t)e_{\ominus\alpha}(t, 0)}{1 + M\mu(t)},$$

where q is given as in (i).

(iii) $\bar{v}(t, U(t)) - (\bar{v}(t, U(t)))^{r/q} \leq 0$, *where q and r are constants given as in* (i) *and* (ii), *respectively.*

Then the trivial solution of SDIVP (4.3) *is exponentially stable on* $\mathbb{T}$.

Proof. In virtue of Theorem 4.12 and our assumptions, we have

$$[\bar{v}(t, U(t))e_M(t, 0)]^{\Delta}$$

$$= \bar{v}^{\Delta}(t, U(t))e_M(\sigma(t), 0) + M\bar{v}(t, U(t))e_M(t, 0)$$

$$\leq (-\lambda_3(t)(\|U(t)\|)^r - L(M \ominus \alpha)(t)e_{\ominus\alpha}(t, 0))e_M(t, 0)$$

$$+ M\bar{v}(t, U(t))e_M(t, 0)$$

$$\leq \left(\frac{-\lambda_3(t)}{\lambda_2^{r/q}(t)} \bar{v}^{r/q}(t, U(t)) + M\bar{v}(t, U(t)) \right.$$

$$\left. - L(M \ominus \alpha)(t)e_{\ominus\alpha}(t, 0) \right) e_M(t, 0)$$

$$\leq \left(M(\bar{v}(t, U(t)) - \bar{v}^{r/q}(t, U(t))) - L(M \ominus \alpha)(t)e_{\ominus\alpha}(t, 0) \right) e_M(t, 0)$$

$$\leq -L(M \ominus \alpha)(t)e_{\ominus\alpha}(t, 0)e_M(t, 0)$$

$$= -L(M \ominus \alpha)(t)e_{M\ominus\alpha}(t, 0).$$

That is,

$$[\bar{v}(t, U(t))e_M(t, 0)]^{\Delta} \leq -L(M \ominus \alpha)(t)e_{M\ominus\alpha}(t, 0) \quad \text{for all } t \in \mathbb{T}.$$

Integrating this inequality from t_0 to t and observing $U_0 = U(t_0)$, we obtain

$$\bar{v}(t, U(t))e_M(t, 0) \le \bar{v}(t_0, U_0)e_M(t_0, 0) - Le_{M \ominus \alpha}(t, 0) + Le_{M \ominus \alpha}(t_0, 0)$$
$$\le \bar{v}(t_0, U_0)e_M(t_0, 0) + Le_{M \ominus \alpha}(t_0, 0)$$
$$\le (\bar{v}(t_0, U_0) + L)e_M(t_0, 0).$$

This yields

$$\bar{v}(t, U(t)) \le (\bar{v}(t_0, U_0) + L)e_M(t_0, 0)e_{\ominus M}(t, 0)$$
$$= (\bar{v}(t_0, U_0) + L)e_{\ominus M}(t, t_0).$$

From this and our assumption, it follows that

$$\|U(t)\| \le \lambda_1^{-1/p}(t) \left((\bar{v}(t_0, U_0) + L)e_{\ominus M}(t, t_0) \right)^{1/p}$$
$$\le \lambda_1^{-1/p}(t_0)((\bar{v}(t_0, U_0) + L)e_{\ominus M}(t, t_0))^{1/p}.$$

This, in view of Definition 4.6(I), yields the desired result. This completes the proof. $\qquad \square$

Remark 4.2. In Theorem 4.14, if $\lambda_i(t) = \lambda_i$, $i = 1, 2, 3$, are positive constants, then the trivial solution of SDIVP (4.3) is uniformly exponentially stable cn $\mathbb{T}$.

Theorem 4.15. *Assume that $\bar{v} \in \mathcal{C}_{rd}^1(\mathbb{T} \times K_c^1, \mathbb{R}^+)$ and $U \in \mathcal{C}_{rd}(\mathbb{T}, K_c^1)$ satisfy the following conditions.*

(i) *There exist constants $k_1 \ge 0$ and $p \ge 0$ such that*

$$k_1\|U(t)\|^p \le \bar{v}(t, U(t)).$$

(ii) *There exist nonnegative constants k_2, L, and positive constants α and β with $0 < \beta \le \min\{k_2, \alpha\}$ such that*

$$\bar{v}^\Delta(t, U(t)) \le \frac{-k_2\bar{v}(t, U(t)) - L(\beta \ominus \alpha)(t)e_{\ominus \alpha}(t, 0)}{1 + \beta\mu(t)}.$$

Then the trivial solution of SDIVP (4.3) is uniformly exponentially stable on $\mathbb{T}$.

Proof. By means of Theorem 4.12 and our assumptions, we have

$$[\bar{v}(t, U(t))e_\beta(t, 0)]^\Delta$$

$$= \bar{v}^\Delta(t, U(t))e_\beta(\sigma(t), 0) + \beta\bar{v}(t, U(t))e_\beta(t, 0)$$

$$\leq (-k_2\bar{v}(t, U(t)) - L(\beta \ominus \alpha)(t)e_{\ominus\alpha}(t, 0))e_\beta(t, 0)$$

$$\quad + \beta\bar{v}(t, U(t))e_\beta(t, 0)$$

$$= (-k_2\bar{v}(t, U(t)) + \beta\bar{v}(t, U(t)) - L(\beta \ominus \alpha)(t)e_{\ominus\alpha}(t, 0)) \, e_\beta(t, 0)$$

$$\leq -L(\beta \ominus \alpha)(t)e_{\ominus\alpha}(t, 0)e_\beta(t, 0)$$

$$= -L(\beta \ominus \alpha)(t)e_{\beta\ominus\alpha}(t, 0).$$

That is,

$$[\bar{v}(t, U(t))e_\beta(t, 0)]^\Delta \leq -L(\beta \ominus \alpha)(t)e_{\beta\ominus\alpha}(t, 0) \quad \text{for all } t \in \mathbb{T}.$$

Integrating this inequality from t_0 to t with $U_0 = U(t_0)$, we obtain

$$\bar{v}(t, U(t))e_\beta(t, 0) \leq \bar{v}(t_0, U_0)e_\beta(t_0, 0) - Le_{\beta\ominus\alpha}(t, 0) + Le_{\beta\ominus\alpha}(t_0, 0)$$

$$\leq \bar{v}(t_0, U_0)e_\beta(t_0, 0) + Le_{\beta\ominus\alpha}(t_0, 0)$$

$$\leq (\bar{v}(t_0, U_0) + L)e_\beta(t_0, 0).$$

This yields

$$\bar{v}(t, U(t)) \leq (\bar{v}(t_0, U_0) + L)e_\beta(t_0, 0)e_{\ominus\beta}(t, 0)$$

$$= (\bar{v}(t_0, U_0) + L)e_{\ominus\beta}(t, t_0).$$

From this and our assumption, it follows that

$$\|U(t)\| \leq k_1^{-1/p}((\bar{v}(t_0, U_0) + L)e_{\ominus\beta}(t, t_0))^{1/p}.$$

This, in view of Definition 4.6(II), yields the desired result. This completes the proof. $\qquad\square$

Theorem 4.16. *Let the assumptions of Theorem 4.14 hold expect that the estimate in* (ii) *is strengthened to*

$$\bar{v}^\Delta(t, U(t)) \leq -\lambda_3(t)\|U(t)\|^r,$$

where λ_3 is nondecreasing positive function. Assume that λ_1 in (i) *is nondecreasing and $\bar{v}(t, \theta) = 0$ for $(t, U(t)) \in \mathbb{T} \times K_c^1$. Then the trivial solution of SDIVP* (4.3) *is exponentially asymptotically stable.*

Proof. Theorem 4.14 guarantees that the trivial solution of SDIVP (4.3) is exponentially stable. Thus, (4.19) holds. From Theorem 1.9 and Bernoulli's inequality, given in Theorem 1.12, it follows that:

$$e_{\ominus M}(t, t_0) = \frac{1}{e_M(t, t_0)}$$

$$\leq \frac{1}{1 + M(t - t_0)}.$$

This yields

$$\lim_{t \to \infty} \varrho(\|U_0\|, t_0)(e_{\ominus M}(t, t_0))^d = 0.$$

From our assumptions, it follows that

$$\bar{v}(t, U(t)) + \int_{t_0}^{t} \lambda_3(s) \|U(s)\|^r \Delta s \leq \bar{v}(t_0, U_0) \quad \text{for all } t \in [t_0, \infty)_{\mathbb{T}}.$$

Note that $\bar{v}(t, U(t)) \geq 0$ and $\bar{v}^{\Delta}(t, U(t)) \leq 0$ guarantee that the function $\bar{v}(t, U(t))$ is decreasing in $t \in \mathbb{T}$. This shows that there is $\beta \in \mathbb{R}$ such that

$$\lim_{t \to \infty} \bar{v}(t, U(t)) = \beta.$$

We prove that $\beta = 0$. On the contrary, suppose $\beta > 0$. Then by means of the fact that $\bar{v}(t, U(t))$ is decreasing, we have

$$\bar{v}(t, U(t)) \geq \beta > 0 \quad \text{for all } t \in [t_0, \infty)_{\mathbb{T}}.$$

Also, from the continuity of $\bar{v}(t, U(t))$ with respect to U and $\bar{v}(t, \theta) = 0$, there is a positive constant ξ such that $\|U(t)\| > \xi$ for each $t \in [t_0, \infty)_{\mathbb{T}}$. This, together with the fact that λ_3 is nondecreasing function positive, yields that

$$\bar{v}(t, U(t)) \leq \bar{v}(t_0, U_0) - \int_{t_0}^{t} \lambda_3(s) \|U(s)\|^r \Delta s$$

$$\leq \bar{v}(t_0, U_0) - \lambda_3(t_0)\xi^r(t - t_0) \quad \text{for all } t \in [t_0, \infty)_{\mathbb{T}}.$$

Let $t \in \mathbb{T}$ be large enough. Then we obtain that $\bar{v}(t, U(t)) < 0$, a contradiction. Hence, $\beta = 0$, that is,

$$\lim_{t \to \infty} \bar{v}(t, U(t)) = 0.$$

Now, we will prove that

$$\lim_{t \to \infty} \|U(t)\| = 0.$$

If this were false, there would exists positive number $\varepsilon_0 > 0$ such that, for $m \in \mathbb{N}$, $\|U(t_m)\| > \varepsilon_0 > 0$ for some $t_m \in \mathbb{T}$ with $t_m \geq m$. From this, combing the fact that λ_1 is positive nondecreasing function, we obtain

$$\bar{v}(t_m, U(t_m)) \geq \lambda_1(t_m)\|U(t_m)\|^p$$

$$\geq \lambda_1(t_m)\varepsilon_0^p$$

$$> 0.$$

This contradicts to the fact

$$\lim_{t \to \infty} \bar{v}(t, U(t)) = 0.$$

Hence,

$$\lim_{t \to \infty} \|U(t)\| = 0$$

which guarantees that the trivial solution of SDIVP (4.3) is exponentially asymptotically stable. This completes the proof. $\square$

Theorem 4.17. *Let the assumptions of Theorem 4.15 hold except that the estimate in* (ii) *is strengthened to*

$$\bar{v}^\Delta(t, U(t)) \leq -k_2 \bar{v}(t, U(t)). \tag{4.20}$$

Assume that $\bar{v}(t, \theta) = 0$ *for* $(t, U(t)) \in \mathbb{T} \times K_c^1$. *Then the trivial solution of SDIVP* (4.3) *is uniformly exponentially asymptotically stable.*

Proof. Theorem 4.15 guarantees that the trivial solution of SDIVP (4.3) is uniformly exponentially stable. Thus, (4.19) holds. From Theorem 1.9 and Bernoulli's inequality, given in Theorem 1.12, it follows that

$$e_{\ominus M}(t, t_0) = \frac{1}{e_M(t, t_0)}$$

$$\leq \frac{1}{1 + M(t - t_0)}.$$

This yields

$$\lim_{t \to \infty} \varrho(\|U_0\|)(e_{\ominus \varepsilon}(t, t_0))^d = 0.$$

In view of our assumptions, it follows that

$$\bar{v}(t, U(t)) + \int_{t_0}^t k_2 \bar{v}(s, U(s)) \Delta s \leq \bar{v}(t_0, U_0) \quad \text{for all } t \in [t_0, \infty)_{\mathbb{T}}.$$

Since $k_1 \|U(t)\|^p \leq \bar{v}(t, U(t))$, we have

$$\bar{v}(t, U(t)) \leq \bar{v}(t_0, U_0) - \int_{t_0}^t k_2 k_1 \|U(s)\|^p \Delta s.$$

Note that $\bar{v}(t, U(t)) \geq 0$ and $\bar{v}^\Delta(t, U(t)) \leq 0$ guarantee that the function $\bar{v}(t, U(t))$ is decreasing in $t \in \mathbb{T}$. This shows that there is $\alpha \in \mathbb{R}$ such that

$$\lim_{t \to \infty} \bar{v}(t, U(t)) = \alpha.$$

We prove that $\alpha = 0$. On the contrary, suppose $\alpha > 0$. Then, by means of the fact that $\bar{v}(t, U(t))$ is decreasing, we have

$$\bar{v}(t, U(t)) \geq \alpha > 0 \quad \text{for all } t \in [t_0, \infty)_{\mathbb{T}}.$$

Repeating the process of the proof of Theorem 4.16, we obtain that $\bar{v}(t, U(t)) < 0$, a contradiction. Hence, $\alpha = 0$, that is,

$$\lim_{t \to \infty} \bar{v}(t, U(t)) = 0.$$

Now, we will prove that

$$\lim_{t \to \infty} \|U(t)\| = 0.$$

If this were false, then there would exists $\varepsilon_0 > 0$ such that, for any $n \in \mathbb{N}$ and for some $t_n \in \mathbb{T}$ with $t_n \geq n$, we have $\|U(t_n)\| > \varepsilon_0 > 0$. Then,

$$\bar{v}(t_n, U(t_n)) \geq k_1 \|U(t_n)\|^p$$
$$\geq k_1 \varepsilon_0^p$$
$$> 0.$$

This contradicts to the fact that

$$\lim_{t \to \infty} \bar{v}(t, U(t)) = 0.$$

Hence,

$$\lim_{t \to \infty} \|U(t)\| = 0,$$

showing that, the trivial solution of SDIVP (4.3) is uniformly exponentially asymptotically stable. This completes the proof. □

Example 4.3. Consider the SDIVP (4.16). Choose the Lyapunov-like function

$$\bar{v}(t, U(t)) = \|U(t)\| \quad \text{for } t \in \mathbb{T}.$$

Then, for $U(t) = U(t, t_0, U_0)$, a solution of SDIVP (4.16) corresponding the initial value $(0, U_0)$, we have

$$\Delta^r \bar{v}(t, U(t))$$

$$= \begin{cases} \dfrac{\|U(\sigma(t))\| - \|U(t)\|}{\mu(t)} & \text{for } \sigma(t) > t \\[4mm] \lim_{s \to t^+} \sup \dfrac{\|U(t) + (s-t)(\ominus U(t))\| - \|U(t)\|}{s-t} \\[4mm] \quad = -\|U(t)\| & \text{for } \sigma(t) = t, \end{cases}$$

and

$$\Delta_r \bar{v}(t, U(t))$$

$$= \begin{cases} \dfrac{\|U(\sigma(t))\| - \|U(t)\|}{\mu(t)} & \text{for } \sigma(t) > t \\[4mm] \lim_{s \to t^+} \inf \dfrac{\|U(t) + (s-t)(\ominus U(t))\| - \|U(t)\|}{s-t} \\[4mm] \quad = -\|U(t)\| & \text{for } \sigma(t) = t. \end{cases}$$

In addition,

$$\frac{\|U(\sigma(t))\| - \|U(t)\|)}{\mu(t)} \le \frac{\|U(\sigma(t) - U(t))\|}{\mu(t)}$$

$$= \|\Delta_g U(t)\|$$

$$= \| \ominus U(t) \|$$

$$= -\frac{1}{1 + \mu(t)} \| U(t) \|$$

and

$$-\| U(t) \| \leq -\frac{1}{1 + \mu(t)} \| U(t) \|.$$

This implies that

$$\bar{v}^{\Delta}(t, U(t)) \leq -\frac{1}{1 + \mu(t)} \| U(t) \| \quad \text{for all } t \in \mathbb{T}.$$

Now, taking $k_1 = k_2 = 1$, $p = 1$, $L = 0$, $\beta = 1$, and $\alpha = 1$, we see that the conditions (i) and (ii) in Theorem 4.15 are satisfied. As a conclusion, SDIVP (4.16) is uniformly exponentially stable. Moreover, (4.20) is satisfied with $k_2 = 1$, from Theorem 4.17, it follows that the trivial solution of SDIVP (4.16) is uniformly exponentially asymptotically stable.

Remark 4.3. In particular, $\mu(t) = 0$ when we choose $\mathbb{T} = \mathbb{R}$. Consequently,

$$\bar{v}^{\Delta}(t, U(t)) = -\| U(t) \|$$

under the assumptions of Example 4.3. Taking $k_1 = k_2 = \frac{1}{2}$, $p = 1$, $L = 0$, $\beta = \frac{1}{2}$, and $\alpha = 1$, we see that the conditions (i) and (ii) in Theorem 4.15 are satisfied. Therefore, (4.16) is a set differential equation in the sense of Hukuhara derivative (see Lakshmikantham *et al.*, 2004, for more detail) and its trivial solution is uniformly exponentially stable.

If $\mathbb{T} = \mathbb{N}$, then $\mu(t) = 1$. So

$$\bar{v}^{\Delta}(t, U(t)) = -\frac{1}{2} \| U(t) \|.$$

Taking $k_1 = k_2 = \frac{1}{2}$, $p = 1$, $L = 0$, $\beta = \frac{1}{4}$, and $\alpha = 2$, we see that the conditions (i) and (ii) in Theorem 4.15 are satisfied. Conclusively, (4.16) is a set difference (see Bhaskar and Shaw, 2004, for more detail) and its trivial solution is uniformly exponentially stable.

4.4.2 *Generalization*

In this section, we discuss the stability of the following SDIVP

$$\Delta_g X(t) = F(t, X), \quad X(t_0) = X_0 = (X_1^0, X_2^0, \ldots, X_m^0)^T \in (K_c^n)^m, \tag{4.21}$$

where $X = (X_1, X_2, \ldots, X_m)^T$ with $m \in \mathbb{N}$ and $X_i \colon \mathbb{T} \to K_c^n$ for $1 \le i \le m$, $t_0 \in \mathbb{T}$ is given and $F \colon \mathbb{T} \times (K_c^n)^m \to (K_c^n)^m$ is a given set-valued function, $(K_c^n)^m = K_c^n \times K_c^n \times \cdots \times K_c^n$ (m times). Moreover, $(K_c^n)^m$ is endowed with the distance as follows

$$D_0[X, Y] = \sum_{i=1}^{m} D[X_i, Y_i]$$

for $X, Y \in (K_c^n)^m$. It can be easily seen that $((K_c^n)^m, D_0)$ is a metric space. Let $F \in \mathcal{C}_{rd}(\mathbb{T} \times (K_c^n)^m), (K_c^n)^m)$ and

$$\mathscr{X} = \{X \in \mathcal{C}_{rd}(\mathbb{T}, (K_c^n)^m) \colon X(t) = X(t, t_0, X_0) \text{ is a solution of}$$

$$\text{SDIVP (4.21)}\}$$

be nonempty. Next, together with SDIVP (4.21), we will consider the matrix-valued function $\tilde{u} \colon \mathbb{T} \times (K_c^n)^m \to \mathbb{R}^{m \times m}$ defined by

$$\tilde{u}(t, X) = [\tilde{u}_{ij}(t, X)], \quad 1 \le i, j \le m, \quad (t, X) \in \mathbb{T} \times (K_c^n)^m, \tag{4.22}$$

with

$$\tilde{u}_{ii} \in \mathcal{C}_{rd}(\mathbb{T} \times (K_c^n)^m, \mathbb{R}^+), \quad i = 1, 2, \ldots, m,$$

and

$$\tilde{u}_{ij} \in \mathcal{C}_{rd}(\mathbb{T} \times (K_c^n)^m, \mathbb{R}), \quad i, j = 1, 2, \ldots, m, \ i \ne j.$$

In the following, we suppose that $\det \tilde{u}(t, \Theta_0) = 0$, where $\det \tilde{u}(t, X)$ stands for the determinant of $\tilde{u}(t, X)$ and Θ_0 stands for the zero element in $(K_c^n)^m$. Let $\mathcal{A}(t, X)$ denote the set consisting of all solutions of the homogeneous linear system

$$\tilde{u}(t, X)a = 0, \quad (t, X) \in \mathbb{T} \times (K_c^n)^m, \quad a \in \mathbb{R}^{+m}.$$

It is clear that system

$$\tilde{u}(t, \Theta_0)a = 0$$

has nontrivial solutions. More precisely, $\mathcal{A}(t, \Theta_0)$ contains at least a nonvanishing vector for $t \in \mathbb{T}$. Based on the matrix-valued function given in (4.22), the scalar function $\bar{v}$ defined as

$$\bar{v}(t, X, a) = a^T \tilde{u}(t, X)a, \qquad (4.23)$$

where $(t, X, a) \in \mathbb{T} \times (K_c^n)^m \times \mathbb{R}^{+m}$, is called a Lyapunov-like function. Clearly, we see that $\bar{v} \in \mathcal{C}_{rd}(\mathbb{T} \times (K_c^n)^m \times \mathbb{R}^{+m}, \mathbb{R}^+)$ and $\bar{v}(t, \Theta_0, a) = 0$ for each $t \in \mathbb{T}$ and each $a \in \mathcal{A}(t, \Theta_0)$. The delta derivative with respect to $t \in \mathbb{T}$ for a Lyapunov-like function $\bar{v}$ in (4.23) is

$$\bar{v}^\Delta(t, X, a) = a^T \tilde{u}^\Delta(t, X)a,$$

where

$$\tilde{u}^\Delta(t, A)$$

$$= \begin{cases} \dfrac{\tilde{u}(\sigma(t), A(\sigma(t))) - \tilde{u}(t, A(t))}{\mu(t)} & \text{for } \sigma(t) > t \\[4mm] \limsup\limits_{s \to t^+} \dfrac{\tilde{u}(t + \mu(t), A + (s - t)F(t, A)) - \tilde{u}(t, A)}{s - t} & \text{for } \sigma(t) = t, \end{cases}$$

for any given $A \in \mathcal{C}_{rd}(\mathbb{T}, (K_c^n)^m)$.

Similar to Theorem 4.12, we have the following result:

Theorem 4.18. *Let* $\bar{v}(t, X, a) = a^T U(t, X)a$ *be a Lyapunov-like function. Then its ur and lr delta derivatives exist. Moreover, for any fixed* $A \in \mathcal{C}_{rd}(\mathbb{T}, (K_c^n)^m)$, $a \in \mathbb{R}^{+m}$, *the delta derivative of* $\bar{v}$ *with respect to* $t \in \mathbb{T}$ *exists and*

$$\bar{v}^\Delta(t, A(t), a) = \Delta^r \bar{v}(t, A(t), a) = \Delta_r \bar{v}(t, A(t), a).$$

Proof. The proof is left as an exercise. $\qquad\square$

Let

$$Q = \left\{ H \colon \mathbb{T} \times (K_c^n)^m \to \mathbb{R}^+ \colon \inf_{X \in (K_c^n)^m} H(t, X) = 0 \right\}$$

and

$$Q_0 = \left\{ H_0 \in Q \colon \inf_X H_0(t, X) = 0 \text{ for } t \in \mathbb{R}^+ \right\}.$$

These two sets Q and Q_0 characterize the current and initial states, respectively, of the set of solutions to SDIVP (4.21). Moreover, these two sets are used to establish some stability conditions, such as (H_0, H)-stable, (H_0, H)-uniformly stable, and (H_0, H)-asymptotically stable, under two different measures based on a class of matrix-valued Lyapunov functions for SDIVP (4.21). We consider two more sets

$$\mathcal{M} = \big\{ H \in \mathcal{C}_{\mathrm{rd}}(\mathbb{T} \times (K_c^n)^m, \mathbb{R}^+) \colon H(t, \cdot) \text{ cannot be a constant with}$$
$$\text{respect to the second variable on } \mathscr{X} \big\}$$

and

$$\mathcal{M}_0 = \{ h_0 \colon h_0 = H_0(t_0, X_0) \text{ for some given } H_0 \in \mathcal{M} \}.$$

These sets $\mathcal{M}$ and $\mathcal{M}_0$ characterize the current and initial states, respectively, of the set $\mathscr{X}$ of solutions to SDIVP (4.21). These sets are used to establish the H-exponential stability, H-exponentially asymptotic stability, H-uniformly exponential stability and H-uniformly exponentially asymptotic stability for the trivial solution of SDIVP (4.21) with the hypothesis that the initial value in SDIVP (4.21) with $t_0 \in \mathbb{T}$ is positive. In addition, the following notations will be used:

$$\Gamma := \{ \gamma \colon [0, \infty) \to [0, \infty) \colon \gamma \text{ is strictly increasing continuous}$$
$$\text{with } \gamma(0) = 0 \}$$

and

$$\Lambda := \{ \lambda \colon [0, \infty) \to [0, \infty) \colon \lambda \text{ is continuous, } \lambda(0) = 0, \text{ and}$$
$$\lambda(s) > 0 \text{ for } s > 0 \}.$$

Definition 4.7. Let $H, H_0 \in \mathcal{M}$, $X \in \mathscr{X}$ and a constant $p \in (0, +\infty)$. Then the trivial solution of SDIVP (4.21) is said to be

(I) H-exponentially stable on $\mathbb{T}$ provided there exist $\eta \in \Gamma$ and a function $\varrho \colon \mathbb{R}^+ \times \mathbb{T} \to \mathbb{R}^+$ such that

$$\eta\left((H(t, X(t)))^p\right) \le \varrho(h_0, t_0)(e_{\ominus M}(t, t_0))^d \text{ for } t \in [t_0, \infty)_{\mathbb{T}},$$
$$(4.24)$$

where $M \in \mathcal{R}_1^+$, $d \in (0, +\infty)$, and $h_0 = H_0(t_0, X_0)$.

(II) H-uniformly exponentially stable provided (I) holds with the function ϱ independent of t_0.

(III) H-exponentially asymptotically stable provided (I) holds, as well as, for any given $\varepsilon > 0$, there is a positive real number T such that

$$\eta\left((H(t, X(t)))^p\right) < \varepsilon \quad \text{for all } t \in [t_0 + T, \infty)_{\mathbb{T}}.$$

(IV) H-uniformly exponentially asymptotically stable provided there is $\eta \in \Gamma$ such that (II) and (III) hold simultaneously.

Theorem 4.19. *Let $\bar{v} \in \mathcal{C}_{\mathrm{rd}}(\mathbb{T} \times (K_c^n)^m \times \mathbb{R}^{+m}, \mathbb{R}^+)$ be a Lyapunov-like function. For $H \in \mathcal{M}$, $X \in \mathcal{X}$, a constant $p > 0$ and a vector $a \in \mathbb{R}^{+m}$, suppose the following conditions hold.*

(i) *there exist functions $\lambda_1, \lambda_2 \in \Lambda$ such that*

$$\lambda_1\left((H(t, X(t)))^p\right) \le \bar{v}(t, X(t), a) \le \lambda_2\left((H(t, X(t)))^p\right)$$

for all $t \in \mathbb{T}$.

(ii) *there exist a nondecreasing continuous function $\lambda_3 \colon \mathbb{R}^+ \to \mathbb{R}$, functions $\gamma \in \Gamma$, $\alpha \in \mathcal{R}_1$, $M \in \mathcal{R}_1^+$ with $\lambda_2(t) \le \gamma(t)$ for $t \in \mathbb{T}^+$ and constants L, r with $r > 0$ such that*

$$\bar{v}^\Delta(t, X(t), a) \le \frac{-\lambda_3\left((H(t, X(t)))^r\right) - L(M \ominus \alpha)(t)e_{\ominus\alpha}(t, 0)}{1 + M\mu(t)},$$

$$\tag{4.25}$$

and

$$M\bar{v}(t, X(t), a) \le \lambda_3\left(\left[\gamma^{-1}(\bar{v}(t, X(t), a))\right]^{r/p}\right)$$

$$+ L(M \ominus \alpha)(t)e_{\ominus\alpha}(t, 0) \tag{4.26}$$

for all $t \in \mathbb{T}$.

Then, the trivial solution of SDIVP (4.21) is H-exponentially stable on $[t_0, \infty)_\mathbb{T}$.

Proof. Let $\bar{\gamma} \in \Gamma$ satisfy

$$\bar{\gamma}((H(t, X(t)))^p) \le \lambda_1((H(t, X(t)))^p)$$

and

$$\lambda_2((H(t, X(t)))^p) \le \gamma((H(t, X(t)))^p).$$

Combining this with the condition (i), we have

$$\bar{\gamma}((H(t, X(t)))^p) \le \lambda_1((H(t, X(t)))^p)$$

$$\le \bar{v}(t, X(t), a)$$

$$\le \lambda_2((H(t, X(t)))^p)$$

$$\le \gamma\left((H(t, X(t)))^p\right).$$

That is,

$$\bar{\gamma}((H(t, X(t)))^p) \leq \gamma ((H(t, X(t)))^p) \qquad (4.27)$$

for all $t \in \mathbb{T}$. To show the desired result, it is sufficient to verify (4.24). The remaining part of the proof is divided into three steps.

Step 1. We first verify that $\bar{v}(t, X(t), a)e_M(t, t_0)$ is nonincreasing in $t \in [t_0, \infty)_{\mathbb{T}}$. Indeed, by means of (4.25), together with Theorem 1.9 (i) and Theorem 4.18, we have

$$[\bar{v}(t, X(t), a)e_M(t, t_0)]^{\Delta}$$

$$= \bar{v}^{\Delta}(t, X(t), a)e_M(\sigma(t), t_0) + M\bar{v}(t, X(t), a)e_M(t, t_0)$$

$$= \left(\bar{v}^{\Delta}(t, X(t), a)e_M(\sigma(t), t) + M\bar{v}(t, X(t), a)\right) e_M(t, t_0)$$

$$\leq \left(\frac{-\lambda_3 ((H(t, X(t)))^r) - L(M \ominus \alpha)(t)e_{\ominus \alpha}(t, 0)}{1 + M\mu(t)} e_M(\sigma(t), t)\right.$$

$$\left. + M\bar{v}(t, X(t), a)\right) \times e_M(t, t_0)$$

$$= \left(-\lambda_3 ((H(t, X(t)))^r) - L(M \ominus \alpha)(t)e_{\ominus \alpha}(t, 0)\right.$$

$$\left. + M\bar{v}(t, X(t), a)\right)e_M(t, t_0).$$

From (4.27), it follows that

$$\left[\gamma^{-1}(\bar{v}(t, X(t), a))\right]^{r/p} \leq (H(t, X(t)))^r .$$

By virtue of the monotonicity of λ_3, we have

$$-\lambda_3 \left(\left[\gamma^{-1}(\bar{v}(t, X(t), a))\right]^{r/p}\right) \geq -\lambda_3((H(t, X(t)))^r).$$

This, combining (4.26) and (4.27), implies that

$$[\bar{v}(t, X(t), a)e_M(t, t_0)]^{\Delta}$$

$$\leq \left(-\lambda_3 ((H(t, X(t)))^r) - L(M \ominus \alpha)(t)e_{\ominus \alpha}(t, 0)\right.$$

$$\left. + M\bar{v}(t, X(t), a)\right)e_M(t, t_0)$$

$$\leq \left(-\lambda_3 \left([\gamma^{-1}(\bar{v}(t, X(t), a))]^{r/p}\right) - L(M \ominus \delta)(t)e_{\ominus \alpha}(t, 0)\right.$$
$$\left. + M\bar{v}(t, X(t), a)\right) e_M(t, t_0)$$
$$\leq 0.$$

Consequently, $\bar{v}(t, X(t), a)e_M(t, t_0)$ is nonincreasing in t, $t \in [t_0, \infty)_{\mathbb{T}}$.

Step 2. For the sake of convenience, let $N > 1$ be a given constant and $u(t_0, X_0) = N\bar{v}(t_0, X_0, a)$. We claim that

$$\bar{v}(t, X(t), a) \leq u(t_0, X_0)e_{\ominus M}(t, t_0) \quad \text{for all } t \in [t_0, \infty)_{\mathbb{T}}. \quad (4.28)$$

Suppose that the inequality (4.28) does not hold. Then, there exists $t \in [t_0, \infty)_{\mathbb{T}}$ such that

$$\bar{v}(t, X(t), a) > u(t_0, X_0)e_{\ominus M}(t, t_0).$$

Set

$$\bar{t} = \inf \left\{ t \in [t_0, \infty)_{\mathbb{T}} : \bar{v}(t, X(t), a) > u(t_0, X_0)e_{\ominus M}(t, t_0) \right\}.$$

Then, from Step 1, it follows that $\bar{t} > t_0$ (otherwise, our claim is achieved). Without any restriction, assume that

$$\bar{v}(\bar{t}, X(\bar{t}), a) \geq u(t_0, X_0)e_{\ominus M}(\bar{t}, t_0) \quad (4.29)$$

and

$$\bar{v}(t, X(t), a) \leq u(t_0, X_0)e_{\ominus M}(t, t_0) \quad \text{for all } t \in [t_0, \bar{t})_{\mathbb{T}}. \quad (4.30)$$

Next, choose $\varphi \in \Gamma$ such that $s < \varphi(s) \leq Ns$ for any $s \geq 0$. Then, we have

$$\varphi\left(\bar{v}(\bar{t}, X(\bar{t}), a)\right) \geq \varphi\left(u(t_0, X_0)e_{\ominus M}(\bar{t}, t_0)\right)$$
$$> u(t_0, X_0)e_{\ominus M}(\bar{t}, t_0)$$

and

$$\varphi\left(\bar{v}(t_0, X_0, a)\right) = \varphi\left(N^{-1}u(t_0, X_0)\right)$$
$$\leq u(t_0, X_0).$$

Note that the set

$$\left\{ t \in [t_0, \bar{t}]_{\mathbb{T}} : \varphi\left(\bar{v}(t, X(t), a)\right) \leq u(t_0, X_0)e_{\ominus M}(t, t_0) \right\}$$

is nonempty, since it includes at least one element t_0. Thus, we define

$$t^* = \sup \{t \in [t_0, \bar{t}]_{\mathbb{T}} \colon \varphi\left(\bar{v}(t, X(t), a)\right) \le u(t_0, X_0)e_{\ominus M}(t, t_0)\}.$$

and we deduce that

$$\varphi\left(\bar{v}(t^*, X(t^*), a)\right) \le u(t_0, X_0)e_{\ominus M}(t^*, t_0) \quad \text{for } t^* \in [t_0, \bar{t})_{\mathbb{T}}, \quad (4.31)$$

and

$$\varphi\left(\bar{v}(t, X(t), a)\right) > u(t_0, X_0)e_{\ominus M}(t, t_0) \quad \text{for } t \in (t^*, \bar{t}]_{\mathbb{T}}. \quad (4.32)$$

Now, Step 1 guarantees that $\bar{v}(t, X(t), a)e_M(t, t_0)$ is nonincreasing in t, $t \in [t^*, \bar{t}]_{\mathbb{T}}$, which implies that

$$\bar{v}(\bar{t}, X(\bar{t}), a)e_M(\bar{t}, t_0) \le \bar{v}(t^*, X(t^*), a)e_M(t^*, t_0).$$

On the other hand, from (4.29) and (4.31), it follows that

$$\begin{aligned}
\bar{v}(\bar{t}, X(\bar{t}), a)e_M(\bar{t}, t_0) &\ge u(t_0, X_0) \\
&\ge \varphi\left(v(t^*, X(t^*), a)\right)e_M(t^*, t_0) \\
&> v(t^*, X(t^*), a)e_M(t^*, t_0).
\end{aligned}$$

This is a contradiction and hence (4.28) is true.

Step 3. Finally, according to the condition (i), we derive

$$\begin{aligned}
\bar{\gamma}\left((H(t, X(t)))^p\right) &\le \bar{v}(t, X(t), a) \\
&\le u(t_0, X_0)e_{\ominus M}(t, t_0)
\end{aligned}$$

That is,

$$\bar{\gamma}\left((H(t, X(t)))^p\right) \le u(t_0, X_0)e_{\ominus M}(t, t_0) \quad \text{for all } t \in [t_0, \infty)_{\mathbb{T}}. \quad (4.33)$$

Let $\eta = \bar{\gamma} \in \Gamma$, and

$$\varrho(h_0, t_0) = u(t_0, X_0) = N\bar{v}(t_0, X_0, a),$$

and

$$h_0 = H_0(t_0, X_0) = v(t_0, X_0, a) \quad \text{with } H_0 \in \mathcal{M}.$$

Consequently, (4.33) guarantees that (4.24) is satisfied. Hence, the trivial solution of SDIVP (4.21) is H-exponentially stable on $[t_0, \infty)_{\mathbb{T}}$. This completes the proof. $\qquad \square$

Corollary 4.1. *Let $\bar{v} \in \mathcal{C}_{\mathrm{rd}}(\mathbb{T} \times (K_c^n)^m \times \mathbb{R}^{+m}, \mathbb{R}^+)$ be a Lyapunov-like function. For $H \leqq \mathcal{M}, X \in \mathscr{X}$, suppose that the following conditions are satisfied.*

(i) *There exist positive functions $\eta_1, \eta_2 \in \Lambda$ and positive constants p, q such that*

$$\eta_1(t)(H(t, X(t)))^p \leq \bar{v}(t, X(t), a) \leq \eta_2(t)(H(t, X(t)))^q$$

for all $t \in \mathbb{T}$.

(ii) *There exist a positive function $\lambda_3 \colon \mathbb{T}^+ \to \mathbb{R}^+$ with $M \in \mathcal{R}_1^+$, where*

$$M =: \inf_{s \geq t_0} \frac{\lambda_3(s)}{[\eta_2(s)]^{\frac{r}{q}}},$$

function $\alpha \in \mathcal{R}_1$, and constants L, r with $r > 0$ such that

$$\bar{v}^\Delta(t, X(t), a) \leq \frac{-\lambda_3(t)(H(t, X(t)))^r - L(M \ominus \delta)(t)e_{\ominus\alpha}(t, 0)}{1 + M\mu(t)}$$

and

$$M\left(\bar{v}(t, X(t), a) - v^{r/q}(t, X(t), a)\right) \leq L(M \ominus \alpha)(t)e_{\ominus\alpha}(t, 0),$$

where q is given as in (i).

Then, the trivial solution of SDIVP (4.21) *is H-exponentially stable on $[t_0, \infty)_{\mathbb{T}}$.*

Proof. In fact, take

$$\lambda_1\left((H(t, X(t)))^p\right) = \eta_1(t)(H(t, X(t)))^p$$

and

$$\lambda_2\left((H(t, X(t)))^p\right) = \eta_2(t)(H(t, X(t)))^q.$$

Then the condition (i) of Theorem 4.19 holds. It is easy to verify that the remaining conditions of Theorem 4.19 are satisfied. In virtue of Theorem 4.19, the trivial solution of SDIVP (4.21) is H-exponentially stable on $[t_0, \infty)_{\mathbb{T}}$. This completes the proof. $\qquad\square$

Corollary 4.2. *Let $Y \in \mathcal{C}^1_{\mathrm{rd}}(\mathbb{T} \times K^1_c, \mathbb{R}^+)$ be a Lyapunov-like function satisfying the following conditions on $\mathbb{T} \times K^1_c$. For $X \in \mathscr{X}$,*

(i) *there exist positive functions $\mu_1, \mu_2 \in \Lambda$ and positive constants p, q such that*

$$\eta_1(t)\|X(t)\|^p \leq Y(t, X(t)) \leq \eta_2(t)\|X(t)\|^q \quad \text{for all } t \in \mathbb{T}.$$

(ii) *there exist a positive function λ_3 on $\mathbb{T}$ with $M \in \mathcal{R}^+_1$, where*

$$M =: \inf_{s \geq t_0} \frac{\lambda_3(s)}{[\eta_2(s)]^{r/q}},$$

function $\alpha \in \mathcal{R}_1$, and constants L, r with $r > 0$ such that

$$Y^\Delta(t, X(t)) \leq \frac{-\lambda_3(t)\|X(t)\|^r - L(M \ominus \alpha)(t)e_{\ominus \alpha}(t, 0)}{1 + M\mu(t)},$$

$$M(Y(t, X(t)) - Y^{r/q}(t, X(t))) \leq L(M \ominus \alpha)(t)e_{\ominus \alpha}(t, 0),$$

where q is given as in (i).

Then, the trivial solution of the following SDIVP

$$\Delta_g X(t) = F(t, X), \quad X(t_0) = X_0 \in K^1_c, \tag{4.34}$$

is H-exponentially stable on $[t_0, \infty)_{\mathbb{T}}$.

Proof. Set $\bar{v}(t, X(t), a) = Y(t, X(t))$, $H(t, X(t)) = \|X(t)\|$, $H_0(t_0, X_0) = \|X_0\|$, and $m = n = 1$. Then we see, from Corollary 4.1, that the trivial solution of (4.34) is H-exponentially stable on $[t_0, \infty)_{\mathbb{T}}$. This completes the proof. □

Remark 4.4. Corollary 4.2 is essentially an extension and improvement of Theorem A.9.

Theorem 4.20. *Let $\bar{v} \in \mathcal{C}_{\mathrm{rd}}(\mathbb{T} \times (K^n_c)^m \times \mathbb{R}^{+m}, \mathbb{R}^+)$ be a Lyapunov-like function and the function $\hat{v} \colon (K^n_c)^m \times \mathbb{R}^{+m} \to \mathbb{R}^+$ satisfy*

$$N\bar{v}(t, X_0, a) \leq \hat{v}(X_0, a) \quad \text{for the constant} \quad N > 1.$$

Moreover, for $H \in \mathcal{M}, X \in \mathscr{X}$, suppose that the following conditions hold.

(i) *There exist constants k_1, $p > 0$ such that*

$$k_1(H(t, X(t)))^p \leq \bar{v}(t, X(t), a) \quad \text{for all } t \in \mathbb{T}.$$

(ii) *There exist constants k_2, L and functions $\beta \in \mathcal{R}_1^+$, $\alpha \in \mathcal{R}_1$ such that*

$$\bar{v}^{\Delta}(t, X(t), a) \leq \frac{-k_2 \bar{v}(t, X(t), a) - L(\beta \ominus \alpha)(t) e_{\ominus \alpha}(t, 0)}{1 + \beta \mu(t)},$$
(4.35)

and

$$(\beta - k_2)\bar{v}(t, X(t), a) \leq L(\beta \ominus \alpha)(t)e_{\ominus \alpha}(t, 0) \quad \text{for all } t \in \mathbb{T}.$$
(4.36)

Then, the trivial solution of SDIVP (4.21) is H-uniformly exponentially stable on $[t_0, \infty)_{\mathbb{T}}$.

Proof. Let $\gamma \in \Gamma$ satisfy $\gamma(s) \leq k_1 s$ for $s \geq 0$. Then, by condition (i), we have

$$\gamma\left((H(t, X(t)))^p\right) \leq k_1(H(t, X(t)))^p$$
$$\leq \bar{v}(t, X(t), a) \quad \text{for all } t \in \mathbb{T}.$$
(4.37)

In the view of Definition 4.7 and inequality (4.37), it suffices to check that

$$\bar{v}(t, X(t), a) \leq N\hat{v}(X_0, a)e_{\ominus \beta}(t, t_0)$$
(4.38)

for $t \in [t_0, \infty)_{\mathbb{T}}$ and fixed $a \in \mathbb{R}^{+m}$. If inequality (4.38) is false, then there exists $t \in [t_0, \infty)_{\mathbb{T}}$ such that

$$\bar{v}(t, X(t), a) > N\hat{v}(X_0, a)e_{\ominus \beta}(t, t_0)$$
$$> \hat{v}(X_0, a)e_{\ominus \beta}(t, t_0),$$

which is possible, since $N > 1$. This implies that

$$\bar{t} = \inf\{t \in [t_0, \infty)_{\mathbb{T}} : \bar{v}(t, X(t), a) > \hat{v}(X_0, a)e_{\ominus \beta}(t, t_0)\}$$

exists. Therefore, we obtain

$$\bar{v}(\bar{t}, X(\bar{t}), a) \geq \hat{v}(X_0, a)e_{\ominus \beta}(\bar{t}, t_0) \quad \text{for } \bar{t} \in (t_0, \infty)_{\mathbb{T}}$$
(4.39)

and

$$\bar{v}(t, X(t), a) \leq \hat{v}(X_0, a)e_{\ominus \beta}(t, t_0) \quad \text{for } t \in [t_0, \bar{t})_{\mathbb{T}}.$$
(4.40)

Again, we consider the function $\varphi \in \Gamma$ with

$$s < \varphi(s) \leq Ns \quad \text{for any } s \geq 0.$$

Then, by (4.39), we have

$$\varphi\left(\bar{v}(\bar{t}, X(\bar{t}), a)\right) \geq \varphi\left(\hat{v}(X_0, a)e_{\ominus\beta}(\bar{t}, t_0)\right)$$
$$> \hat{v}(X_0, a)e_{\ominus\beta}(\bar{t}, t_0).$$

However, we observe that

$$\varphi\left(\bar{v}(t_0, X_0, a)\right) \leq \varphi\left(N^{-1}\hat{v}(X_0, a)\right)$$
$$\leq \hat{v}(X_0, a).$$

This implies that the set

$$\{t \in [t_0, \bar{t}]_{\mathbb{T}} : \varphi\left(\bar{v}(t, X(t), a)\right) \leq \hat{v}(X_0, a)e_{\ominus\beta}(t, t_0)\}$$

is nonempty, and hence its supremum, denoted by t^*, exists. Thus, we have

$$\varphi\left(\bar{v}(t^*, X(t^*), a)\right) \leq \hat{v}(X_0, a)e_{\ominus\beta}(t^*, t_0) \quad \text{for all } t^* \in [t_0, \bar{t})_{\mathbb{T}} \quad (4.41)$$

and

$$\varphi\left(\bar{v}(t, X(t), a)\right) > \hat{v}(X_0, a)e_{\ominus\beta}(t, t_0) \quad \text{for all } t \in (t^*, \bar{t}]_{\mathbb{T}}. \quad (4.42)$$

On the other hand, similar to the proof of Step 1 in Theorem 4.19, we obtain that $\bar{v}(t, X(t), a)e_\beta(t, t_0)$ is nonincreasing in $t \in [t^*, \bar{t}]$, which infers that

$$\bar{v}(t^*, X(t^*), a)e_\beta(t^*, t_0) \geq \bar{v}(\bar{t}, X(\bar{t}), a)e_\beta(\bar{t}, t_0)$$
$$\geq \hat{v}(X_0, a)$$
$$\geq \varphi\left(\bar{v}(t^*, X(t^*), a)\right) e_\beta(t^*, t_0)$$
$$> v(t^*, X(t^*), a)e_\beta(t^*, t_0),$$

a contradiction. Consequently, (4.38) holds and hence the trivial solution of SDIVP (4.21) is H-uniformly exponentially stable on $[t_0, \infty)_{\mathbb{T}}$. This completes the proof. $\qquad \square$

Theorem 4.21. *Let the assumptions of Theorem 4.19 hold, except that function M is changed into a constant and the estimate in (4.25) is strengthened to*

$$\bar{v}^{\Delta}(t, X(t), a) \leq -\lambda_3 (H(t, X(t)))^r, \qquad (4.43)$$

where $\lambda_3 \colon \mathbb{R}^+ \to \mathbb{R}^+$ is a positive nondecreasing function with

$$\lim_{t \to \infty} \lambda_3(t) = \infty$$

and constant $r > 0$. Then, the trivial solution of SDIVP (4.21) is H-exponentially asymptotically stable on $[t_0, \infty)_{\mathbb{T}}$.

Proof. Theorem 4.19 guarantees that the trivial solution of SDIVP (4.21) is H-exponentially stable, that is, (4.24) holds. By Theorem 1.9(vii) and Bernoulli's inequality, given in Theorem 1.12, we obtain

$$e_{\ominus M}(t, t_0) = \frac{1}{e_M(t, t_0)}$$

$$\leq \frac{1}{1 + M(t - t_0)}.$$

This means that

$$\lim_{t \to \infty} \varrho(h_0, t_0)(e_{\ominus M}(t, t_0))^d = 0.$$

Now, integrating both sides of (4.43) from t_0 to t, yields

$$\bar{v}(t, X(t), a) \leq \bar{v}(t_0, X_0, a) - \int_{t_0}^{t} \lambda_3 (((H(s, X(s)))^r)) \Delta_g s \qquad (4.44)$$

for all $t \in [t_0, \infty)_{\mathbb{T}}$. We observe that (4.43) guarantees that $\bar{v}(t, X(t), a)$ is nonincreasing in $t \in \mathbb{T}$. Also, $\bar{v}(t, X(t), a) \geq 0$. From the monotone convergence theorem, it follows that, there exists $\beta \in \mathbb{R}$ with

$$\lim_{t \to \infty} \bar{v}(t, X(t), a) = \beta.$$

We prove that $\beta = 0$. Suppose that this were false. Then, we have $\beta > 0$ and by virtue of the monotonicity of $\bar{v}(t, X(t), a)$, we obtain

$$\bar{v}(t, X(t), a) \geq \beta > 0 \quad \text{for all } t \in [t_0, \infty)_{\mathbb{T}}.$$

On the other hand, since λ_3 is nondecreasing positive and

$$\lim_{t\to\infty} \lambda_3(t) = \infty,$$

we see that

$$\int_{t_0}^t \lambda_3((H(s, X(s)))^r)\Delta s > \bar{v}(t_0, X_0, a),$$

when $t \in \mathbb{T}$ is sufficiently large. Using this in (4.44), we obtain $\bar{v}(t, X(t), a) < 0$, which contradicts $\bar{v}(t, X(t), a) \geq 0$. Hence, $\beta = 0$, that is,

$$\lim_{t\to\infty} \bar{v}(t, X(t), a) = 0.$$

Next, we prove that

$$\lim_{t\to\infty} H(t, X(t)) = 0.$$

If this is not true, then there exists $\varepsilon_0 > 0$ such that $H(t_m, X(t_m)) > \varepsilon_0$ for $m \in \mathbb{N}$ and some $t_m \in \mathbb{T}$ with $t_m \geq m$. From this, combining with (4.27), we have, for $\bar{\gamma} \in \Gamma$,

$$\bar{v}(t_m, X(t_m), a) \geq \bar{\gamma}\left((H(t_m, X(t_m)))^p\right)$$
$$\geq \bar{\gamma}\left(\varepsilon_0^p\right)$$
$$> \bar{\gamma}(0) = 0.$$

This contradicts to the fact that

$$\lim_{t\to\infty} \bar{v}(t, X(t), a) = 0.$$

Hence,

$$\lim_{t\to\infty} H(t, X(t)) = 0.$$

By virtue of the definition of $\bar{\gamma} \in \Gamma$, we obtain

$$\lim_{t\to\infty} \bar{\gamma}\left((H(t, X(t)))^p\right) = 0.$$

This guarantees that for given $\varepsilon > 0$, there exists a positive real number T such that

$$\bar{\gamma}\left((H(t, X(t)))^p\right) < \varepsilon \quad \text{for all } t \in [t_0 + T, \infty)_{\mathbb{T}}.$$

Thus, the trivial solution of SDIVP (4.21) is H-exponentially asymptotically stable. This completes the proof. $\square$

Theorem 4.22. *Let the assumptions of Theorem 4.20 hold except that function β changes into a constant and the estimate in (4.35) is strengthened to*

$$\bar{v}^\Delta(t, X(t), a) \le -k_2 v(t, X(t), a). \tag{4.45}$$

Assume that $\bar{v}(t, \Theta_0, a) = 0$ for each $t \in \mathbb{T}$ and each $a \in \mathcal{A}(t, \Theta_0)$. Then, the trivial solution of SDIVP (4.21) is H-uniformly exponentially asymptotically stable on $[t_0, \infty)_{\mathbb{T}}$.

Proof. Theorem 4.20 guarantees that the trivial solution of (4.24) is H-uniformly exponential stable on $[t_0, \infty)_{\mathbb{T}}$. As an analogous to the proof of Theorem 4.21, we have

$$\lim_{t \to \infty} \varrho(h_0, t_0)(e_{\ominus\beta}(t, t_0))^q = 0$$

and

$$\lim_{t \to \infty} \gamma\left((H(t, X(t)))^p\right) = 0.$$

Thus, the trivial solution of SDIVP (4.21) is H-uniformly exponentially asymptotically stable on $[t_0, \infty)_{\mathbb{T}}$. This completes the proof. $\square$

For the sake of convenience, in the following examples, we consider $\mathbb{T}^+$ instead of $\mathbb{T}$ as the working platform.

Example 4.4. Consider the set dynamic initial value problem

$$\Delta_g X = \ominus\zeta(t)X, \quad X(0) = X_0 \in (K_c^n)^m, \quad t \in \mathbb{T}^+, \tag{4.46}$$

where $\zeta \in \mathcal{C}_{\mathrm{rd}}(\mathbb{T}^+, \mathbb{R}^+) \cap \mathcal{R}_1^+$ is nondecreasing and satisfies that $w \le \zeta(t)$ on $\mathbb{T}^+$ with the constant $w > 0$. From Lemma A.7, it follows that SDIVP (4.46) has a unique solution $X : \mathbb{T}^+ \to (K_c^n)^m$ given by

$$X(t) = X_0 e_{\ominus\zeta}(t, 0).$$

Choose

$$\bar{v}(t, X(t), a) = H(t, X(t)) = \|X(t)\| \quad \text{for } t \in \mathbb{T}^+.$$

Next, taking $\lambda_1(s) = s$, $\lambda_2(s) = s$ for $s \ge 0$, and $p = 1$, we see that condition (i) of Theorem 4.19 is satisfied. Now, in order to verify condition (ii), we first calculate the delta derivative of $\bar{v}(t, X(t), a)$. By Theorem 1.9(v), we obtain

$$\bar{v}^\Delta(t, X(t), a) = \|X_0\| \ominus \zeta(t) e_{\ominus\zeta}(t, 0)$$

$$= -\frac{\zeta(t)}{1 + \mu(t)\zeta(t)} \|X_0\| e_{\ominus\zeta}(t, 0)$$

$$\leq -\frac{w}{1 + w\mu(t)} \bar{v}(t, X(t), a).$$

Next, we take

$$M(t) = \delta(t) = w, \quad r = 1, \quad L = 1, \quad \lambda_3(s) = ws, \quad \gamma(s) = s$$

for all $t \in \mathbb{T}^+$. Then, the condition (ii) of Theorem 4.19 is satisfied. Consequently, the trivial solution of SDIVP (4.46) is H-exponentially stable.

Example 4.5. Consider the set dynamic initial value problem

$$\Delta_g X(t) = -\eta(t)X(\sigma(t)) + F(t), \quad X(0) = X_0 \in (K_c^n)^m, \quad t \in \mathbb{T}^+,$$
$$(4.47)$$

where $\eta \in \mathcal{C}(\mathbb{T}^+, \mathbb{R}^+) \cap \mathcal{R}_1^+$ is nondecreasing and satisfies that $w \leq \eta(t)$ on $\mathbb{T}^+$ with the constant $w > 0$ and $F \in \mathcal{C}_{\mathrm{rd}}(\mathbb{T}^+, (K_c^n)^m)$ satisfies

$$\int_0^t e_{\ominus\eta}(t, s)\|F(s)\|\Delta s \leq \|X_0\| e_{\ominus\eta}(t, 0) \quad \text{for } t \in \mathbb{T}^+.$$

From Lemma A.8, it follows that SDIVP (4.47) has a unique solution $X \colon \mathbb{T}^+ \to (K_c^n)^m$ given by

$$X(t) = X_0 e_{\ominus\eta}(t, 0) + \int_0^t e_{\ominus\eta}(t, s)F(s)\Delta_g s.$$

Thus,

$$\|X(t)\| = \left\| X_0 e_{\ominus\eta}(t, 0) + \int_0^t e_{\ominus\eta}(t, s)F(s)\Delta_g s \right\|$$

$$\geq \|X_0 e_{\ominus\eta}(t, 0)\| - \left\| \int_0^t e_{\ominus\eta}(t, s)F(s)\Delta_g s \right\|$$

$$\geq \|X_0\| e_{\ominus\eta}(t, 0) - \int_0^t e_{\ominus\eta}(t, s)\|F(s)\|\Delta s,$$

i.e.,

$$\|X(t)\| \geq \|X_0\|e_{\ominus\eta}(t,0) - \int_0^t e_{\ominus\eta}(t,s)\|F(s)\|\Delta s. \qquad (4.48)$$

Let

$$\bar{v}(t,X(t),a) = H(t,X(t)) = \|X_0\|e_{\ominus\eta}(t,0) - \int_0^t e_{\ominus\eta}(t,s)\|F(s)\|\Delta s \qquad (4.49)$$

for $t \in \mathbb{T}^+$ and $a \in \mathbb{R}^{+m}$. Then our hypothesis guarantees that $\bar{v}(t,X(t),a) \geq 0$. Next, we verify that the conditions in Theorem 4.19 are satisfied. Taking $\lambda_1(s) = \lambda_2(s) = s$ for $s \geq 0$, and $p = 1$, we see that the condition (i) of Theorem 4.19 is satisfied. In order to verify condition (ii), we first calculate the delta derivative of $\bar{v}(t,X(t),a)$. By Theorem 1.9 (v), we obtain

$$\bar{v}^\Delta(t,X(t),a) = \left[\|X_0\|e_{\ominus\eta}(t,0) - \int_0^t e_{\ominus\eta}(t,s)\|F(s)\|\Delta s \right]^\Delta$$

$$= \|X_0\| \ominus \eta e_{\ominus\eta}(t,0) - e_{\ominus\eta}(\sigma(t),t)\|F(\sigma(t))\|$$

$$\quad - \int_0^t \ominus\eta e_{\ominus\eta}(t,s)\|F(s)\|\Delta s$$

$$= -\frac{\eta(t)}{1+\mu(t)\eta(t)}\|X_0\|e_{\ominus\eta}(t,0) - \frac{1}{1+\mu(t)\eta(t)}\|F(\sigma(t))\|$$

$$\quad + \frac{\eta(t)}{1+\mu(t)\eta(t)} \int_0^t e_{\ominus\eta}(t,s)\|F(s)\|\Delta s$$

$$= \frac{1}{1+\mu(t)\eta(t)} \left[-\eta(t)\|X_0\|e_{\ominus\eta}(t,0) - \|F(\sigma(t))\| \right.$$

$$\quad \left. + \eta(t) \int_0^t e_{\ominus\eta}(t,s)\|F(s)\|\Delta s \right]$$

$$\leq \frac{-\eta(t)}{1+\mu(t)\eta(t)} \left[\|X_0\|e_{\ominus\eta}(t,0) - \int_0^t e_{\ominus\eta}(t,s)\|F(s)\|\Delta s \right]$$

Keeping in mind (4.49), we can write

$$\bar{v}^\Delta(t,X(t),a) \leq \frac{-w}{1+w\mu(t)}\bar{v}(t,X(t),a). \qquad (4.50)$$

Next, we take

$$M(t) = \delta(t) = w, \quad r = 1, \quad L = 1, \quad \lambda_3(s) = ws, \quad \gamma(s) = s.$$

Thus, the condition (ii) of Theorem 4.19 is satisfied. Consequently, the trivial solution of SDIVP (4.47) is H-exponentially stable.

Finally, we see that all conditions in Theorem 4.20 are satisfied provided we choose $k_1 = \frac{1}{2}$ and $\beta(t) = k_2 = w$. Therefore, Theorem 4.20 guarantees that the trivial solution of SDIVP (4.47) is H-uniformly exponentially stable.

Example 4.6. Consider the SDIVP (4.47) under the hypotheses of Example 4.5. Let $\bar{v}(t, X(t), a)$ be given as in (4.49). It has been shown that the conditions (i) and (ii) of Theorem 4.19 are satisfied by taking $M = w$ (a positive constant) in the computation of Example 4.5. Let

$$\lambda_3(s) = \frac{w}{1 + w\mu(t)} s.$$

Then λ_3 satisfies the hypothesis of Theorem 4.21 and the inequality (4.43) holds by (4.50). Now, Theorem 4.21 guarantees that the trivial solution of SDIVP (4.47) is H-exponentially asymptotically stable. Let

$$\varsigma = \max_{t \in \mathbb{T}^+} \mu(t).$$

To check the conditions of Theorem 4.22, let us take constants $k_1 = \frac{1}{2}$, k_2, β, and the function α as follows:

$$k_2 = \beta = \alpha(t) = \begin{cases} \dfrac{w}{2(1+w\varsigma)} & \text{if } \varsigma < \infty \\ 0 & \text{if } \varsigma = \infty \end{cases}$$

for all $t \in \mathbb{T}^+$. Thus, if $\varsigma < \infty$, then a computation similar to that in the proof of Example 4.5 yields

$$\bar{v}^\Delta(t, X(t), a) \leq \frac{-w}{1 + w\mu(t)} v(t, X(t), a)$$

$$\leq \frac{-w}{1 + w\varsigma} \bar{v}(t, X(t), a)$$

$$< \frac{-w}{2(1 + w\varsigma)} \bar{v}(t, X(t), a)$$

$$= -k_2 \bar{v}(t, X(t), a)$$

for all $t \in \mathbb{T}^+$. If $\varsigma = \infty$, then we have

$$\bar{v}^{\Delta}(t, X(t), a) \leq \frac{-w}{1 + w\mu(t)} \bar{v}(t, X(t), a) \leq 0$$
$$= -k_2 \bar{v}(t, X(t), a)$$

for all $t \in \mathbb{T}^+$. Consequently, (4.45) holds. From (4.48) and (4.49), it follows that that $\bar{v}(t, \Theta_0, a) = 0$ for each $t \in \mathbb{T}^+$ and each $a \in \mathcal{A}(t, \Theta_0)$. Moreover, all the conditions in Theorem 4.20 are clearly satisfied under our assumptions. Now, Theorem 4.22 guarantees that the trivial solution of SDIVP (4.47) is H-uniformly exponentially asymptotically stable.

Chapter 5

Applications to Fuzzy Dynamic Equations

In this chapter, as an application of the theory developed in Chapter 2, a class of linear impulsive fuzzy dynamic equations on time scales is considered by using the generalized differentiability concept on time scales. Some novel criteria and general forms of solutions are established for such models whose significance lies in proposing the possibility to get unifying forms of solutions for discrete and continuous dynamic systems under uncertainty and to unify corresponding problems in the framework of fuzzy dynamic equations on time scales. Finally, illustrative examples are included to show the applicability of our results.

5.1 Background

From now on, unless otherwise mentioned, the matrix-valued functions under consideration are always assumed to belong to $\mathcal{R}_n$.

In what follows, we introduce the necessary definitions and notations for fuzzy numbers on time scales which are the extension of the corresponding concepts in $\mathbb{R}$ (see, for example, Daomond and Kloeden (1994)). Let $\mathbb{T}_f$ denotes the class of fuzzy subsets of $\mathbb{T}$ satisfying the following properties. That is, if $u \in \mathbb{T}_f$, then $u \colon \mathbb{T} \to [0, 1]$ and

(f1) u is normal, i.e., there exists $s_0 \in \mathbb{T}$ such that $u(s_0) = 1$;
(f2) u is a convex fuzzy set on $\mathbb{T}$, that is, for all $t \in [0, 1]$, $a, b \in \mathbb{T}$, satisfying $ta + (1 - t)b \in \mathbb{T}$, we have

$$u(ta + (1 - t)b) \geq \min\{u(a), u(b)\};$$

(f3) u is upper semicontinuous on $\mathbb{T}$; and

(f4) The set $[u]^0 = \overline{\{s \in \mathbb{T} \colon u(s) > 0\}}$ is compact, where $\overline{A}$ denotes the closure of A in $\mathbb{R}$.

Then $\mathbb{T}_f$ is called the space of fuzzy numbers. Obviously, $\mathbb{T} \subset \mathbb{T}_f$. Here $\mathbb{T} \subset \mathbb{T}_f$ is understood as

$$\mathbb{T} = \{\chi_{\{x\}} \colon x \text{ is an element of time scale}\},$$

where χ is the characteristic function. For $0 < \alpha \leq 1$, define the α-set

$$[u]^\alpha = \{s \in \mathbb{T} \colon u(s) \geq \alpha\}$$

and

$$[u]^0 = \overline{\{s \in \mathbb{T} \colon u(s) > 0\}} \cap \mathbb{T}.$$

From (f1)–(f4), it follows that the α-level set $[u]^\alpha$ is a nonempty compact interval of $\mathbb{T}$ for all $0 \leq \alpha \leq 1$ provided u belongs to $\mathbb{T}_f$. That is, $[u]^\alpha$ is an intersection of a nonempty compact interval of $\mathbb{R}$ and $\mathbb{T}$. The notation

$$[u]^\alpha = [\underline{u}^\alpha, \overline{u}^\alpha] \cap \mathbb{T} =: [\underline{u}^\alpha, \overline{u}^\alpha] \quad \text{with } \underline{u}^\alpha, \overline{u}^\alpha \in \mathbb{T}$$

denotes explicitly the α-level set of u on $\mathbb{T}$. We refer to $\underline{u}$ and $\overline{u}$ as the lower branch and the upper branch of u, respectively. For $u \in \mathbb{T}_f$, we define the length of u as

$$\mathrm{diam}(u) = \overline{u}^\alpha - \underline{u}^\alpha.$$

For $\alpha \in [0,1]$ and $u, v \in \mathbb{T}_f$, the sum $u + v$ is defined as

$$[u + v]^\alpha = [u]^\alpha + [v]^\alpha$$

and the product λu is defined as

$$[\lambda u]^\alpha = \lambda [u]^\alpha, \quad \text{for } \lambda \in \mathbb{R},$$

where $[u]^\alpha + [v]^\alpha$ denotes the usual addition of two subsets $[u]^\alpha$ and $[v]^\alpha$ of $\mathbb{T}$ and $\lambda [u]^\alpha$ denotes the usual product between scalar λ and a subset $[u]^\alpha$ of $\mathbb{T}$. The metric structure is given by the Hausdorff distance $D_f \colon \mathbb{T}_f \times \mathbb{T}_f \to \mathbb{R}^+$,

$$D_f[u, v] = \sup_{\alpha \in [0,1]} \max\left\{|\underline{u}^\alpha - \underline{v}^\alpha|, |\overline{u}^\alpha - \overline{v}^\alpha|\right\}.$$

Note that $(\mathbb{T}_f, D_f)$ is a complete metric space. We here adopt H-difference and have the following remarks.

(1) Denote $\hat{0} \in \mathbb{T}_f$ as a neutral element with respect to "+" provided

$$u + \hat{0} = \hat{0} + u = u \quad \text{for all } u \in \mathbb{T}_f.$$

We have $\hat{0} = \chi_{\{0\}}$ provided $0 \in \mathbb{T}$.

(2) $(\lambda + \mu)u = (\lambda u) + (\mu u)$ with $\lambda\mu > 0$.

(3) $\lambda(u + v) = (\lambda u) + (\lambda v)$.

The notion of strongly generalized differentiability on $\mathbb{R}$ was introduced by Bede and Gal (2005) and studied further in Bede *et al.* (2007) and Li *et al.* (2012). Motivated by these works, Hong *et al.* (2020) introduced generalized differentiability on time scale $\mathbb{T}$ which is due to Vasavi *et al.* (2016) and can be regarded as a generalization of Δ_H-differentiability introduced by Hong (2009).

Definition 5.1. Let $F \colon \mathbb{T} \to \mathbb{T}_f$ be a fuzzy set-valued function, $t \in \mathbb{T}$, and a neighborhood $U_\mathbb{T}$ of t be defined by $U_\mathbb{T} = (t - \delta, t + \delta) \cap \mathbb{T}$ for some $\delta > 0$. Then,

(1) F is said to be delta right differentiable at t provided there is an element $\Delta_+ F(t)$ of $\mathbb{T}_f$ and, for given $\varepsilon > 0$, there is a neighborhood $U_\mathbb{T}$ of t such that

(c^1) either the H-difference $F(t + h) -_H F(\sigma(t))$ exists and

$$D[F(t + h) -_H F(\sigma(t)), \Delta_+ F(t)(h - \mu(t))] \le \varepsilon|h - \mu(t)|,$$

(c^2) or $F(\sigma(t)) -_H F(t + h)$ exists and

$$D[F(\sigma(t)) -_H F(t + h), \Delta_+ F(t)(\mu(t) - h)] \le \varepsilon|h - \mu(t)|$$

for all $t + h \in U_\mathbb{T}$ with $0 \le h < \delta$. Moreover, $\Delta_+ F(t)$ is called the delta right derivative of F at t.

(2) F is said to be delta left differentiable at t provided there exists an element $\Delta_- F(t)$ of $\mathbb{T}_f$ and, for given $\varepsilon > 0$, there is a neighborhood $U_\mathbb{T}$ of t such that

(c_1) either the H-difference $F(t - h) -_H F(\sigma(t))$ exists and

$$D[F(t - h) -_H F(\sigma(t)), \Delta_- F(t)(-h - \mu(t))] \le \varepsilon|h + \mu(t)|,$$

or $F(\sigma(t)) -_H F(t - h)$ exists and

(c_2)

$$D[F(\sigma(t)) -_H F(t - h), \Delta_- F(t)(h + \mu(t))] \le \varepsilon|\mu(t) + h|$$

for all $t - h \in U_{\mathbb{T}}$ with $0 \leq h < \delta$. Moreover, $\Delta_- F(t)$ is called the delta left derivative of F at t.

(3) F is said to be (i)-differentiable at t provided (c^1) and (c_2) hold and

$$\Delta_+ F(t) = \Delta_- F(t) := \Delta_1 F(t),$$

and is said to be (ii)-differentiable at t provided (c^2) and (c_1) hold and

$$\Delta_+ F(t) = \Delta_- F(t) := \Delta_2 F(t).$$

(4) F is said to be delta differentiable at t provided F is both delta left and delta right differentiable at t and $\Delta_- F(t) = \Delta_+ F(t)$. The common element $\Delta_+ F(t)$ or $\Delta_- F(t)$ is said to be the Δ_H-derivative of F at t and is denoted by $\Delta_H F(t)$. We say that F is Δ_H-differentiable at t provided its Δ_H-derivative exists at t. Moreover, we say F is Δ_H-differentiable on $\mathbb{T}$ provided its Δ_H-derivative exists at each $t \in \mathbb{T}$. The fuzzy set-valued function $\Delta_H F : \mathbb{T} \to \mathbb{T}_f$ is then called the Δ_H-derivative of F on $\mathbb{T}$.

Note 5.1. In view of Definition 5.1, if F is (i)-differentiable at t, then

$$\Delta_1 F(t) = \Delta_H F(t)$$

and if F is (ii)-differentiable at t, then

$$\Delta_2 F(t) = \Delta_H F(t).$$

We shall discuss some important properties of Δ_H-derivative in the sense of Definition 5.1. The majority of demonstrations are similar to the previous theory discussed in this book.

Lemma 5.1. *Let* $a \colon \mathbb{T} \to \mathbb{R}$ *be delta differentiable and* $G \colon \mathbb{T} \to \mathbb{T}_f$ *be* Δ_H-*differentiable. Then we have the following:*

(a) *If* $a(\sigma(t))a^\Delta(t) > 0$ *and* G *is* (i)-*differentiable, then* aG *is* (i)-*differentiable, and*

$$\Delta_H(aG)(t) = a^\Delta(t)G(t) + a(\sigma(t))\Delta_H G(t).$$

(b) *If $a(\sigma(t))a^{\Delta}(t) < 0$ and G is (ii)-differentiable, then aG is (ii)-differentiable, and*

$$\Delta_H(aG)(t) = a^{\Delta}(t)G(t) + a(\sigma(t))\Delta_H G(t).$$

(c) *If $a(\sigma(t))a^{\Delta}(t) > 0$, G is (ii)-differentiable, and the H-differences*

$$(aG)(\sigma(t)) -_H (aG)(t+h) \quad and \quad (aG)(t-h) -_H (aG)(\sigma(t))$$

exist, then aG is (ii)-differentiable, and

$$\Delta_H(aG)(t) = a(\sigma(t))\Delta_H G(t) -_H (-a^{\Delta}(t))G(t).$$

Proof. We shall go into details regarding discussion of the cases (b) and (c) while the proof of (a) is similar to that of (b).

(b) For $t \in \mathbb{T}$ and $\varepsilon > 0$, there is a neighborhood $U_{\mathbb{T}}$ of t and $\delta > 0$ such that

$$D[G(\sigma(t)) -_H G(t+h), \Delta_H G(t)(\mu(t) - h)] \le \varepsilon|h - \mu(t)| \quad (5.1)$$

and

$$|a(\sigma(t)) - a(t+h) - a^{\Delta}(t)(\mu(t) - h)| \le \varepsilon|h - \mu(t)| \quad (5.2)$$

for $0 \le h < \delta$ with $t + h \in U_{\mathbb{T}}$. On the other hand, we have

$D[a(\sigma(t))G(\sigma(t)) -_H a(t+h)G(t+h),$

$\quad (a^{\Delta}(t)G(t) + a(\sigma(t))\Delta_H G(t))(\mu(t) - h)]$

$\le D[a(\sigma(t))G(\sigma(t)) -_H a(\sigma(t))G(t+h),$

$\qquad a(\sigma(t))\Delta_H G(t)(\mu(t) - h)]$

$\quad + D[a(\sigma(t))G(t+h) -_H a(t+h)G(t+h), a^{\Delta}(t)G(t)(\mu(t) - h)]$

$\le D[G(\sigma(t)) -_H G(t+h), \Delta_H G(t)(\mu(t) - h)]|a(\sigma(t))|$

$\quad + D[a(\sigma(t))G(t+h) -_H a(t+h)G(t+h), a^{\Delta}(t)G(t)(\mu(t) - h)]$

$\le D[G(\sigma(t)) -_H G(t+h), \Delta_H G(t)(\mu(t) - h)]|a(\sigma(t))|$

$\quad + D[a(\sigma(t))G(t+h), a(t+h)G(t+h)$

$\quad + a^{\Delta}(t)G(t+h)(\mu(t) - h)]$

$\quad + D[\{0\}, a^{\Delta}(t)G(t)(\mu(t) - h) -_H a^{\Delta}(t)G(t+h)(\mu(t) - h)].$

That is, for $t \in \mathbb{T}$ and for $0 \leq h < \delta$ with $t + h \in U_{\mathbb{T}}$,

$$D[a(\sigma(t))G(\sigma(t)) -_H a(t+h)G(t+h),$$
$$(a^\Delta(t)G(t) + a(\sigma(t))\Delta_H G(t))(\mu(t) - h)]$$
$$\leq D[G(\sigma(t)) -_H G(t+h), \Delta_H G(t)(\mu(t) - h)]|a(\sigma(t))|$$
$$+ D[a(\sigma(t))G(t+h), a(t+h)G(t+h) + a^\Delta(t)G(t+h)(\mu(t) - h)]$$
$$+ D[\{0\}, a^\Delta(t)G(t)(\mu(t) - h) -_H a^\Delta(t)G(t+h)(\mu(t) - h)].$$
$$(5.3)$$

Note that $a(\sigma(t))$ and $a(t + h)$ have the same sign for sufficiently small $h > 0$. Moreover, $a(\sigma(t))a^\Delta(t) < 0$ and $\mu(t) - h \leq 0$ imply that $a^\Delta(t)(\mu(t) - h)$ and $a(\sigma(t))$ have the same sign. Hence,

$$a(t+h)G(t+h) + a^\Delta(t)G(t+h)(\mu(t) - h)$$
$$= [a(t-h) + a^\Delta(t)(\mu(t) - h)]G(t+h)$$

and

$$D[a(\sigma(t))G(t+h), a(t+h)G(t+h) + a^\Delta(t)G(t+h)(\mu(t) - h)]$$
$$= |a(\sigma(t)) - (a(t+h) + a^\Delta(t)(\mu(t) - h))|\|G(t+h)\|.$$

Substituting this into the inequality (5.3), we have

$$D[a(\sigma(t))G(\sigma(t)) -_H a(t+h)G(t+h),$$
$$(a^\Delta(t)G(t) + a(\sigma(t))\Delta_H G(t))(\mu(t) - h)]$$
$$\leq D[G(\sigma(t)) -_H G(t+h), \Delta_H G(t)(\mu(t) - h)]|a(\sigma(t))|$$
$$+ |a(\sigma(t)) - (a(t+h) + a^\Delta(t)(\mu(t) - h))|\|G(t+h)\|$$
$$+ D[G(t), G(t+h)]|a^\Delta(t)(\mu(t) - h)|.$$

In view of this and the continuity of G, together with the inequalities (5.1) and (5.2), we see that aG satisfies the first inequality of Definition 5.1-(c_2). We can similarly check the second inequality of Definition 5.1-(c_1). Consequently, the desired conclusion arrives.

(c) As in case (b), the inequalities (5.1) and (5.2) are valid. Moreover, under the hypothesis of (c), we see that $-a^\Delta(t)(\mu(t) - h)$, $a(\sigma(t))$, and $a(t + h)$ have the same sign. Therefore, for $0 \leq h < \delta$ with $t + h \in U_{\mathbb{T}}$, we have

$$D[a(\sigma(t))G(\sigma(t)) -_H a(t + h)G(t + h),$$
$$\quad (a(\sigma(t))\Delta_H G(t) -_H (-a^{\Delta}(t))G(t))(\mu(t) - h)]$$
$$\leq D[a(\sigma(t))G(\sigma(t)) -_H a(\sigma(t))G(t + h), a(\sigma(t))\Delta_H G(t)(\mu(t) - h)]$$
$$\quad + D[a(\sigma(t))G(t + h) -_H a(t + h)G(t + h),$$
$$\hat{0} -_H (-a^{\Delta}(t)G(t)(\mu(t) - h))]$$
$$\leq D[G(\sigma(t)) -_H G(t + h), \Delta_H G(t)(\mu(t) - h)]|a(\sigma(t))|$$
$$\quad + D[a(\sigma(t))G(t + h), a(t + h)G(t + h)$$
$$\quad + (-a^{\Delta}(t))G(t + h)(\mu(t) - h)]$$
$$\quad + D[\hat{0}, -_H(-a^{\Delta}(t))G(t + h)(\mu(t) - h)$$
$$\quad -_H (-a^{\Delta}(t)G(t)(\mu(t) - h))]$$
$$\leq D[G(\sigma(t)) -_H G(t + h), \Delta_H G(t)(\mu(t) - h)]|a(\sigma(t))|$$
$$\quad + |a(\sigma(t)) - (a(t + h) + a^{\Delta}(t)(\mu(t) - h))|\|G(t + h)\|$$
$$\quad + D[G(t), G(t + h)]|a^{\Delta}(t)(\mu(t) - h)|.$$

As in case (b), we arrive at the desired result. This completes the proof. $\qquad\square$

Theorem 5.1. *Let* $F \colon \mathbb{T} \to \mathbb{T}_f$ *and denote* $[F(t)]^{\alpha} = [\underline{F}^{\alpha}(t), \overline{F}^{\alpha}(t)]$ *for each* $\alpha \in [0, 1]$.

(i) *If* F *is* (i)-*differentiable, then* $\underline{F}^{\alpha}$ *and* $\overline{F}^{\alpha}$ *are delta differentiable functions and*

$$[\Delta_H F(t)]^{\alpha} = [(\underline{F}^{\alpha})^{\Delta}(t), (\overline{F}^{\alpha})^{\Delta}(t)].$$

(ii) *If* F *is* (ii)-*differentiable, then* $\underline{F}^{\alpha}$ *and* $\overline{F}^{\alpha}$ *are delta differentiable functions and we have*

$$[\Delta_H F(t)]^{\alpha} = [(\overline{F}^{\alpha})^{\Delta}(t), (\underline{F}^{\alpha})^{\Delta}(t)].$$

Proof. We only check (ii) since the demonstration of (i) is similar. If t is right-scattered, then by Proposition 2.3(II), we have

$$[\Delta_H F(t)]^\alpha = \left[\frac{F(\sigma(t)) -_g F(t)}{-\mu(t)}\right]^\alpha$$

$$= -\frac{1}{\mu(t)}[F(\sigma(t)) -_g F(t)]^\alpha$$

$$= -\frac{1}{\mu(t)}[\underline{F}(\sigma(t)) - \underline{F}(t), \overline{F}(\sigma(t)) - \overline{F}(t)]$$

$$= \left[\frac{\overline{F}(\sigma(t)) - \overline{F}(t)}{\mu(t)}, \frac{\underline{F}(\sigma(t)) - \underline{F}(t)}{\mu(t)}\right]$$

$$= [(\overline{F}^\alpha)^\Delta(t), (\underline{F}^\alpha)^\Delta(t)].$$

If t is right-dense, then by Proposition 2.3(III), for $h < 0$, we have

$$[F(t+h) -_g F(t)]^\alpha = \left[\underline{F}^\alpha(t+h) -_g \underline{F}^\alpha(t), \overline{F}^\alpha(t+h) -_g \overline{F}^\alpha(t)\right]$$

and, multiplying by $\frac{1}{h}$, we get

$$\left[\frac{F(t+h) -_g F(t)}{h}\right]^\alpha = \frac{1}{h}\left[\underline{F}^\alpha(t+h) -_g \underline{F}^\alpha(t), \overline{F}^\alpha(t+h) -_g \overline{F}^\alpha(t)\right]$$

$$= \left[\frac{\overline{F}^\alpha(t+h) -_g \overline{F}^\alpha(t)}{h}, \frac{\underline{F}^\alpha(t+h) -_g \underline{F}^\alpha(t)}{h}\right]$$

Similarly, for $h > 0$, we obtain

$$\left[\frac{F(t) -_g F(t-h)}{h}\right]^\alpha = \left[\frac{\overline{F}^\alpha(t) -_g \overline{F}^\alpha(t-h)}{h}, \frac{\underline{F}^\alpha(t) -_g \underline{F}^\alpha(t-h)}{h}\right]$$

Now, passing to the limit as h approaches to zero, we get

$$[\Delta_H F(t)]^\alpha = [(\overline{F}^\alpha)^\Delta(t), (\underline{F}^\alpha)^\Delta(t)].$$

This completes the proof. □

Let $\mathbb{J} \subset \mathbb{T}$. A function $f \colon \mathbb{J} \to \mathbb{R}$ is called an integrable selector of the fuzzy-number-valued function $F \colon \mathbb{J} \to \mathbb{T}_f$ provided f is delta integrable and $f(t) \in [F(t)]^\alpha$ for all $t \in \mathbb{J}$ with $\alpha \in [0, 1]$. The function

F is called Δ_H-integrable on $\mathbb{J}$ provided it has at least an integrable selector. The Δ_H-integral of F, denoted by

$$\int_{\mathbb{J}} F(s)\Delta_H s,$$

is defined levelwise by

$$\left[\int_{\mathbb{J}} F(s)\Delta_H s\right]^\alpha = \int_{\mathbb{J}} [F(s)]^\alpha \Delta_g s$$

$$= \left\{ \int_{\mathbb{J}} f(s)\Delta s \colon f \text{ is an integrable selector of } F \text{ on } \mathbb{J} \right\}.$$

For the fuzzy version of the fundamental properties of calculus in the sense of (i)-differentiability, we refer to the analogue of set-valued functions discussed by Hong (2009) and Hong *et al.* (2014) and in the sense of (ii)-differentiability, we present the following results which are similar to those proposed in Chalco-Cano and Román-Flores (2008) and Khastan *et al.* (2011).

Theorem 5.2. *Suppose $F\colon \mathbb{T} \to \mathbb{T}_f$ is Δ_H-integrable. Then the integral*

$$\mathscr{F}(t) = \int_{t_0}^t F(s)\Delta_H s \tag{5.4}$$

is (i)-differentiable and $\Delta_H \mathscr{F}(t) = F(t)$. If $F \in \mathcal{C}_{\mathrm{rd}}(\mathbb{T}, \mathbb{T}_f)$ and the function $\mathscr{F}$, defined in (5.4), is (ii)-differentiable, then $\Delta_H \mathscr{F}(t) = F(t)$ for all $t \in \mathbb{T}$.

Remark 5.1. In general, function $\mathscr{F}$, defined in (5.4), is not (ii)-differentiable. Indeed, if it is (ii)-differentiable, then the length of the support decreases in t. But $\mathscr{F}$ has increasing length of the support whenever F is fuzzy non real-valued function. Also, a (ii)-differentiable function needs to have decreasing length of support, which is a contradiction.

The following theorem is useful whose proof is similar to the corresponding proof for the Δ_g case given in Proposition A.2.

Theorem 5.3. *Let* $\mathbb{J} = [t_0, T] \cap \mathbb{T}$ *with* $t_0, T \in \mathbb{T}$. *Then, we have the following:*

(i) *Let* $F \in \mathcal{C}_{\mathrm{rd}}(\mathbb{J}, \mathbb{T}_f)$ *and define*

$$U(t) = \gamma -_H \int_{t_0}^t -F(s)\Delta_H s$$

for $t \in \mathbb{J}$, *where* $\gamma \in \mathbb{T}_f$ *is such that the above H-difference exists for* $t \in \mathbb{J}$. *Then,* U *is* (ii)-*differentiable and*

$$\Delta_H U(t) = F(t) \quad for \, t \in \mathbb{J}.$$

(ii) *Let* F *be* (ii)-*differentiable and* $\Delta_H F$ *be integrable on* $\mathbb{T}^+$. *Then, for each* $t \in \mathbb{T}^+$ *with* $t \geq t_0 \in \mathbb{T}^+$, *we have*

$$F(t) = F(t_0) -_H \int_{t_0}^t -\Delta_H F(\tau)\Delta_H \tau.$$

(iii) $\int_{t_0}^T F(s)\Delta_H s = \int_{t_0}^t F(s)\Delta_H s + \int_t^T F(s) s \Delta_H s$ *for any* $t \in \mathbb{J}$.

(iv) $\int_t^t F(s)\Delta_H s = \theta$ *for any* $t \in \mathbb{J}$.

(v) $\int_t^{\sigma(t)} F(s)\Delta_H s = \mu(t)F(t)$ *for* $t \in \mathbb{J}$.

(vi) *Let* F *be* (i)-*differentiable on* $\mathbb{T}^+$. *Then,*

$$\mathscr{F}(t) = \gamma + \int_{t_0}^t F(s)\Delta_H s,$$

where $\gamma \in \mathbb{T}_f$, *is* (i)-*differentiable and* $\Delta_H \mathscr{F}(t) = F(t)$.

(vii) *If* $F, G \in \mathcal{C}_{\mathrm{rd}}(\mathbb{T}, \mathbb{T}_f)$, *then* $D[F(\cdot), G(\cdot)] \colon \mathbb{J} \to \mathbb{R}^+$ *is delta integrable and*

$$D\left[\int_{t_0}^T F(s)\Delta_H s, \int_{t_0}^T G(s)\Delta_H s\right] \leq \int_{t_0}^T D[F(s), G(s)]\Delta s.$$

5.2 General Forms of Solutions for Linear Impulsive Fuzzy Dynamic Equations

Let $\mathbb{T}^+ = \{t \in \mathbb{T} \colon t \geq 0\}$,

$$\mathbb{J} = \left\{ t_k \in \mathbb{T}^+ \colon 0 \leq t_0 < t_1 < t_2 < \cdots < t_k < \cdots , \right.$$

$$\left. \text{and} \quad \lim_{k \to \infty} t_k = \infty \right\},$$

$\mathbb{J}^- = [0, t_0] \cap \mathbb{T}^+$ and $\mathbb{J}_k = (t_k, t_{k+1}] \cap \mathbb{T}^+$ for $k \in \mathbb{N}_0$. For $k \in \mathbb{N}$, the right limit of x at t_k, denoted by $x_{t_k^+}$, is given by

$$x_{t_k^+} = \begin{cases} x(t_k^+) & \text{if } t_k \text{ is right-dense} \\ x(\sigma(t_k)) & \text{if } t_k \text{ is right-scattered.} \end{cases}$$

We emphasize the following notations

$$PC = \left\{ U \colon \mathbb{T}^+ \to \mathbb{T}_f \colon U \in \mathcal{C}_{\mathrm{rd}}(\mathbb{J}_k, \mathbb{T}_f) \text{ and } \lim_{t \to t_k^+} U(t) = U(t_k^+) \text{ exists} \right.$$

$$\left. \text{for } k \in \mathbb{N} \right\}.$$

$BC = \{ U \in PC \colon \|U(t)\| = D(U(t), \hat{0}) \text{ is bounded in } \mathbb{T}^+ \}.$

$PC^1 = \{ U \in BC \colon U \text{ is } \Delta_H\text{-differentiable in each interval } (t_{k-1}, t_k)_{\mathbb{T}} \}.$

It is clear that (BC, D_0) is a complete metric space when it is endowed with the distance

$$D_0[U, V] = \sup_{t \in \mathbb{T}^+} D[U(t), V(t)].$$

Consider the first-order linear impulsive fuzzy dynamic equation (LIFDE)

$$\begin{cases} \Delta_H U(t) = r(t)U(t) + F(t) & \text{for } t \in \mathbb{T}^+ \setminus \mathbb{J} \\ U_{t_k^+} = L_k U(t_k) & \text{for } t_k \in \mathbb{J}, \ k \in \mathbb{N}_0 \\ U(0) = U_0 \in \mathbb{T}_f & \text{for } t_0 \in \mathbb{T}^+, \end{cases} \quad (5.5)$$

where $r\colon \mathbb{T}^+ \to \mathbb{R}$, $F \in PC$ and $L_k\colon PC^1 \to PC^1$ is a continuous linear operator, i.e., for any $v, w \in \mathbb{T}_f$ and $a, b \in \mathbb{R}$, we have

$$L_k(av \pm_H bw) = aL_k(v) \pm_H bL_k(w)$$

whenever the corresponding H-differences exist.

Definition 5.2. Let $U \in PC^1$ be a fuzzy-number-valued function such that $\Delta_H U$ exists at every point $t \in \mathbb{T}^+ \setminus \mathbb{J}$. A solution U of LIFDE (5.5) is called the (i)-solution provided U and $\Delta_H U$ exist in the case of (i)-differential and satisfy all equations in (5.5). The definition that U is a (ii)-solution of (5.5) is similar.

To explore the existence of solutions to the LIFDE (5.5), we need the following essential preliminaries. In virtue of Theorem 1.10, the dynamic initial value problem

$$\begin{cases} u^\Delta(t) = A(t)u(t) + f(t) & \text{for } t \in \mathbb{T}, \\ u(0) = v_0 \end{cases}$$

has a unique solution $v\colon \mathbb{T} \to \mathbb{R}^n$ given by

$$v(t) = e_A(t, 0)v_0 + \int_0^t e_A(t, \sigma(s))f(s)\Delta s,$$

where $A \in \mathcal{R}_n$ and $w\colon \mathbb{T} \to \mathbb{R}^n$ is rd-continuous.

Let $A \in \mathcal{R}_n$, $f\colon \mathbb{T}^+ \to \mathbb{R}^n$ be an rd-continuous function such that

$$\lim_{t \to t_k^+} f(t) = f(t_k^+)$$

exists for $k \in \mathbb{N}$ and let L_k be a continuous linear operator defined on $\mathbb{R}^n$ for $k \in \mathbb{N}$. By an analogous of the proof of the above result, it is proved in Tikare and Tisdell (2020) that the linear impulsive dynamic equation

$$\begin{cases} u^\Delta(t) = A(t)u(t) + f(t) & \text{for } t \in \mathbb{J}_k, \ k \in \mathbb{N}_0 \\ u_{t_k^+} = \Phi_k = L_k(u(t_k)) & \text{for } t_k \in \mathbb{J}, \ k \in \mathbb{N}_0 \\ u(t_k) = u_{k-1}(t_k) & \text{for } t_k \in \mathbb{J}, \ k \in \mathbb{N}, \end{cases} \tag{5.6}$$

has a unique solution

$$\begin{cases} u_k(t) = e_A(t, t_k^+)\Phi_k + \int_{t_k}^t e_A(t, \sigma(\tau))f(\tau)\Delta\tau & \text{for } t \in \mathbb{J}_k \\ u_k(t_k) = u_{k-1}(t_k) & \text{for } t_k \in \mathbb{J}, \end{cases} \quad (5.7)$$

for $k \in \mathbb{N}_0$, where $u_{0-1} = v(t_0)$. Thus, we obtain that the linear impulsive dynamic equation

$$\begin{cases} u^\Delta(t) = A(t)u(t) + f(t) & \text{for } t \in \mathbb{T}^+ \setminus \mathbb{J} \\ u_{t_k^+} = \Phi_k = L_k(u(t_k)) & \text{for } t_k \in \mathbb{J}, \ k \in \mathbb{N}_0 \\ u(0) = v_0, \end{cases} \quad (5.8)$$

has a unique solution $w = w(v_0)$ on $\mathbb{T}^+$ which is left continuous on $\mathbb{T}^+$ and defined by

$$w(t) = \begin{cases} v(t) & \text{for } t \in \mathbb{J}^- \\ u_0(t) & \text{for } t \in \mathbb{J}_0 \\ \vdots & \vdots \\ u_k(t) & \text{for } t \in \mathbb{J}_k \\ \vdots & \vdots \end{cases} \quad (5.9)$$

We study the LIFDE (5.5) in three cases $r(t) < 0$, $r(t) > 0$, and $r(t) = 0$ for $t \in \mathbb{T}^+$, where r is a function given in LIFDE (5.5). We first observe that the hyperbolic functions proposed by Bohner and Peterson (2001) can be extended as follows:

$$\cosh_r(t, s) = \frac{e_r(t, s) + e_{-r}(t, s)}{2} \quad \text{and}$$

$$\sinh_r(t, s) = \frac{e_r(t, s) - e_{-r}(t, s)}{2}.$$

Let $E_r(t, s) = e_r(t, s)e_{-r}(t, s)$. Obviously, for all $t, s \in \mathbb{T}$, the hyperbolic functions possess the following important properties:

(p1) $\cosh_r(s, s) = 1$ and $\sinh_r(s, s) = 0$;
(p2)

$$\cosh_r^{\Delta_t}(t, s) = r \sinh_r(t, s)$$

and

$$\sinh_r^{\Delta t}(t, s) = r \cosh_r(t, s),$$

where $\alpha^{\Delta t}(t, s)$ means the delta derivative of α with respect to the variable t;

(p3)

$$\cosh_r(s, t) = \frac{1}{E_r(t, s)} \cosh_r(t, s)$$

and

$$\sinh_r(s, t) = -\frac{1}{E_r(t, s)} \sinh_r(t, s);$$

(p4)

$$\sinh_r(t, s) + \cosh_r(t, s) = e_r(t, s),$$

$$\cosh_r(t, s) - \sinh_r(t, s) = e_{-r}(t, s),$$

and

$$\cosh_r^2(t, s) - \sinh_r^2(t, s) = E_r(t, s).$$

We now state and prove the main results concerning solutions of LIFDE (5.5).

Theorem 5.4. *Suppose that* $r \in \mathcal{R}_1^+$ *satisfies* $r(t) < 0$ *for all* $t \in \mathbb{T}^+$. *Then,*

(i) *the LIFDE* (5.5) *has the* (i)-*solution on* $\mathbb{T}^+$ *given by*

$$U(t) = \begin{cases} V(t) & \text{for } t \in \mathbb{J}^- \\ U_k(t) & \text{for } t \in \mathbb{J}_k, \ k \in \mathbb{N}_0, \end{cases} \tag{5.10}$$

where

$$V(t) = \cosh_r(t, 0)$$

$$\times \left\{ U_0 + \int_0^t F(\tau) \left[\frac{\cosh_r(\sigma(\tau), 0)}{E_r(\sigma(\tau), 0)} -_H \frac{\sinh_r(\sigma(\tau), 0)}{E_r(\sigma(\tau), 0)} \right] \Delta_H \tau \right\}$$

$$+ \sinh_r(t, 0)$$

$$\times \left\{ U_0 + \int_0^t F(\tau) \left[\frac{\cosh_r(\sigma(\tau), 0)}{E_r(\sigma(\tau), 0)} -_H \frac{\sinh_r(\sigma(\tau), 0)}{E_r(\sigma(\tau), 0)} \right] \Delta_H \tau \right\},$$

and

$$U_k(t) = \cosh_r(t, t_k^+) \Big\{ L_k U_k(t_k)$$

$$+ \int_{t_k}^t F(\tau) \left[\frac{\cosh_r(\sigma(\tau), t_k)}{E_r(\sigma(\tau), t_k)} - H \frac{\sinh_r(\sigma(\tau), t_k)}{E_r(\sigma(\tau), t_k)} \right] \Delta_H \tau \Big\}$$

$$+ \sinh_r(t, t_k^+) \Big\{ L_k U_k(t_k)$$

$$+ \int_{t_k}^t F(\tau) \left[\frac{\cosh_r(\sigma(\tau), t_k)}{E_r(\sigma(\tau), t_k)} - H \frac{\sinh_r(\sigma(\tau), t_k)}{E_r(\sigma(\tau), t_k)} \right] \Delta_H \tau \Big\}$$

provided the H-differences exist and

$$U_0(t_0) = V(t_0), \quad U_k(t_k) = U_{k-1}(t_k) \quad \text{for } k \in \mathbb{N}_0.$$

(ii) *The LIFDE (5.5) has the (ii)-solution on $\mathbb{T}^+$ is given by*

$$U(t) = \begin{cases} V(t) & \text{for } t \in \mathbb{J}^- \\ U_k(t) & \text{for } t \in \mathbb{J}_k, \ k \in \mathbb{N}_0, \end{cases} \tag{5.11}$$

where

$$V(t) = e_r(t, 0) \left[U_0 -_H \int_0^t -F(\tau) e_{\ominus r}(\sigma(\tau), 0) \Delta_H \tau \right]$$

and

$$U_k(t) = e_r(t, t_k^+) \left[L_k U_k(t_k) -_H \int_{t_k}^t -F(\tau) e_{\ominus r}(\sigma(\tau), t_k) \Delta_H \tau \right]$$

provided the H-differences exist and

$$U_0(t_0) = V(t_0), \quad U_k(t_k) = U_{k-1}(t_k) \quad \text{for } k \in \mathbb{N}_0.$$

Proof. Theorem 5.1 shows how to translate the LIFDE (5.5) into a system of ordinary dynamic equations, that is, if $r(t) < 0$ with $t \in \mathbb{T}^+$ and U is (i)-differentiable, then

$$[\Delta_H U(t)]^\alpha = [(\underline{U}^\alpha)^\Delta(t), (\overline{U}^\alpha)^\Delta(t)]$$

with $[U]^\alpha = [\underline{U}^\alpha, \overline{U}^\alpha]$ for all $\alpha \in [0,1]$, and (5.5) is transformed into the following impulsive system of dynamic equations

$$
\begin{cases}
(\underline{U}^\alpha)^\Delta(t) = r(t)\overline{U}^\alpha(t) + \underline{F}^\alpha(t) & \text{for } t \in \mathbb{T}^+ \setminus \mathbb{J} \\
(\overline{U}^\alpha)^\Delta(t) = r(t)\underline{U}^\alpha(t) + \overline{F}^\alpha(t) & \text{for } t \in \mathbb{T}^+ \setminus \mathbb{J} \\
(\underline{U}^\alpha)_{t_k^+} = L_k \underline{U}^\alpha(t_k) & \text{for } t_k \in \mathbb{J}, \ k \in \mathbb{N}_0 \\
(\overline{U}^\alpha)_{t_k^+} = L_k \overline{U}^\alpha(t_k) & \text{for } t_k \in \mathbb{J}, \ k \in \mathbb{N}_0 \\
\underline{U}^\alpha(0) = (\underline{U}^\alpha)_0 \\
\overline{U}^\alpha(0) = (\overline{U}^\alpha)_0,
\end{cases}
\tag{5.12}
$$

where $[F(t)]^\alpha = [\underline{F}^\alpha(t), \overline{F}^\alpha(t)]$ for $\alpha \in [0,1]$. To solve the system (5.12), we translate it into the system (5.8) with

$$
u(t) = \begin{pmatrix} \underline{U}^\alpha(t) \\ \overline{U}^\alpha(t) \end{pmatrix} \quad \text{for } \alpha \in [0,1], \ t \in \mathbb{T}^+ \setminus \mathbb{J} \tag{5.13}
$$

$$
u_{t_k^+} = \begin{pmatrix} (\underline{U}^\alpha)_{t_k^+} \\ (\overline{U}^\alpha)_{t_k^+} \end{pmatrix} \quad \text{for } \alpha \in [0,1], \ t \in \mathbb{T}^+ \tag{5.14}
$$

$$
f(t) = \begin{pmatrix} \underline{F}^\alpha(t) \\ \overline{F}^\alpha(t) \end{pmatrix} \quad \text{for } \alpha \in [0,1], \ t \in \mathbb{T}^+ \setminus \mathbb{J} \tag{5.15}
$$

$$
v_0 = \begin{pmatrix} (\underline{U}^\alpha)_0 \\ (\overline{U}^\alpha)_0 \end{pmatrix} \quad \text{for } \alpha \in [0,1], \tag{5.16}
$$

$$
A(t) = \begin{pmatrix} 0 & r(t) \\ r(t) & 0 \end{pmatrix} \quad \text{for } t \in \mathbb{T}^+ \setminus \mathbb{J}. \tag{5.17}
$$

Similarly, if U is (ii)-differentiable, then

$$
[\Delta_H U(t)]^\alpha = [(\overline{U}^\alpha)^\Delta(t), (\underline{U}^\alpha)^\Delta(t)]
$$

and (5.5) is translated into (5.8) with u, $u_{t_k^+}$, v_0 given as in (5.13), (5.14), (5.16), respectively, and

$$
A(t) = \begin{pmatrix} r(t) & 0 \\ 0 & r(t) \end{pmatrix}, \tag{5.18}
$$

$$f(t) = \begin{pmatrix} \overline{F}^{\alpha}(t) \\ \underline{F}^{\alpha}(t) \end{pmatrix}. \tag{5.19}$$

(i) Under the case of the (i)-differentiability, we see that A given by (5.17) belongs to $\mathcal{R}_n$. From the fact $r \in \mathcal{R}_1^+$, it follows that the matrix

$$I + \mu(t)A(t) = \begin{pmatrix} 1 & \mu(t)r(t) \\ \mu(t)r(t) & 1 \end{pmatrix}$$

is invertible for each $t \in \mathbb{T}^+$, that is, $A \in \mathcal{R}$. Moreover, we have

$$e_A(t,s) = \begin{pmatrix} \cosh_r(t,s) & \sinh_r(t,s) \\ \sinh_r(t,s) & \cosh_r(t,s) \end{pmatrix}, \quad s,t \in \mathbb{T}^+,$$

with A given in (5.17).

Now, substituting this matrix exponential function for $e_A(t,t_k^+)$ and $e_A(t,\sigma(\tau))$ into (5.7), we obtain

$$\begin{cases} u_k(t) &= \begin{pmatrix} \cosh_r(t,t_k^+) & \sinh_r(t,t_k^+) \\ \sinh_r(t,t_k^+) & \cosh_r(t,t_k^+) \end{pmatrix} \Phi_k \\ &\quad + \int_{t_k}^t \begin{pmatrix} \cosh_r(t,\sigma(\tau)) & \sinh_r(t,\sigma(\tau)) \\ \sinh_r(t,\sigma(\tau)) & \cosh_r(t,\sigma(\tau)) \end{pmatrix} f(\tau)\Delta\tau \quad \text{for } t \in \mathbb{J}_k \\ u_k(t_k) &= u_{k-1}(t_k) \quad \text{for } t_k \in \mathbb{T}^+, \end{cases}$$

where $u_{0-1}(t_0) = v(t_0)$. Since $e_A(t,\sigma(\tau)) = e_A(t,t_k)e_A(t_k,\sigma(\tau))$, we have

$$u_k(t) = \begin{pmatrix} \cosh_r(t,t_k^+) & \sinh_r(t,t_k^+) \\ \sinh_r(t,t_k^+) & \cosh_r(t,t_k^+) \end{pmatrix}$$

$$\times \left(\Phi_k + \int_{t_k}^t \begin{pmatrix} \cosh_r(t_k,\sigma(\tau)) & \sinh_r(t_k,\sigma(\tau)) \\ \sinh_r(t_k,\sigma(\tau)) & \cosh_r(t_k,\sigma(\tau)) \end{pmatrix} f(\tau)\Delta\tau \right).$$

Therefore, for $t \in \mathbb{J}_k$ with $k \in \mathbb{N}_0$, we have

$$
\begin{pmatrix} \underline{U}^\alpha(t) \\ \overline{U}^\alpha(t) \end{pmatrix} = \begin{pmatrix} \underline{U_k}^\alpha(t) \\ \overline{U_k}^\alpha(t) \end{pmatrix} = \begin{pmatrix} \cosh_r(t, t_k^+) & \sinh_r(t, t_k^+) \\ \sinh_r(t, t_k^+) & \cosh_r(t, t_k^+) \end{pmatrix}
$$

$$
\times \left[\begin{pmatrix} L_k \underline{U_k}^\alpha(t_k) \\ L_k \overline{U_k}^\alpha(t_k) \end{pmatrix} \right.
$$

$$
\left. + \int_{t_k}^t \begin{pmatrix} \cosh_r(t_k, \sigma(\tau)) & \sinh_r(t_k, \sigma(\tau)) \\ \sinh_r(t_k, \sigma(\tau)) & \cosh_r(t_k, \sigma(\tau)) \end{pmatrix} \begin{pmatrix} \underline{F}^\alpha(\tau) \\ \overline{F}^\alpha(\tau) \end{pmatrix} \Delta_H \tau \right]
$$

$$
= \begin{pmatrix} \cosh_r(t, t_k^+) & \sinh_r(t, t_k^+) \\ \sinh_r(t, t_k^+) & \cosh_r(t, t_k^+) \end{pmatrix}
$$

$$
\times \begin{pmatrix} L_k \underline{U_k}^\alpha(t_k) + \int_{t_k}^t (\underline{F}^\alpha(\tau) \cosh_r(t_k, \sigma(\tau)) \\ \quad + \overline{F}^\alpha(\tau) \sinh_r(t_k, \sigma(\tau))) \, \Delta_H \tau \\ L_k \overline{U_k}^\alpha(t_k) + \int_{t_k}^t (\underline{F}^\alpha(\tau) \sinh_r(t_k, \sigma(\tau)) \\ \quad + \overline{F}^\alpha(\tau) \cosh_r(t_k, \sigma(\tau))) \, \Delta_H \tau \end{pmatrix}.
$$

Then, keeping in mind the property (p3), the solution of corresponding system of linear impulsive dynamic equation is given by

$$
[U(t)]^\alpha = \left[\underline{U}^\alpha(t), \overline{U}^\alpha(t) \right],
$$

where

$$
\underline{U}^\alpha(t) = \underline{U_k}^\alpha(t) = \cosh_r(t, t_k^+)
$$

$$
\times \left\{ L_k \underline{U_k}^\alpha(t_k) + \int_{t_k}^t \left(\underline{F}^\alpha(\tau) \frac{\cosh_r(\sigma(\tau), t_k)}{E_r(\sigma(\tau), t_k)} \right.\right.
$$

$$
\left.\left. - \overline{F}^\alpha(\tau) \frac{\sinh_r(\sigma(\tau), t_k)}{E_r(\sigma(\tau), t_k)} \right) \Delta_H \tau \right\}
$$

$$
+ \sinh_r(t, t_k^+) \left\{ L_k \overline{U_k}^\alpha(t_k) + \int_{t_k}^t \left(-\underline{F}^\alpha(\tau) \frac{\sinh_r(\sigma(\tau), t_k)}{E_r(\sigma(\tau), t_k)} \right.\right.
$$

$$
\left.\left. + \overline{F}^\alpha(\tau) \frac{\cosh_r(\sigma(\tau), t_k)}{E_r(\sigma(\tau), t_k)} \right) \Delta_H \tau \right\},
$$

and

$$\overline{U}^\alpha(t) = \overline{U_k}^\alpha(t) = \sinh_r(t, t_k^+)$$

$$\times \left\{ L_k \underline{U_k}^\alpha(t_k) + \int_{t_k}^t \left(\underline{F}^\alpha(\tau) \frac{\cosh_r(\sigma(\tau), t_k)}{E_r(\sigma(\tau), t_k)} \right. \right.$$

$$\left. \left. - \overline{F}^\alpha(\tau) \frac{\sinh_r(\sigma(\tau), t_k)}{E_r(\sigma(\tau), t_k)} \right) \Delta_H \tau \right\}$$

$$+ \cosh_r(t, t_k^+) \left\{ L_k \overline{U_k}^\alpha(t_k) + \int_{t_k}^t \left(-\underline{F}^\alpha(\tau) \frac{\sinh_r(\sigma(\tau), t_k)}{E_r(\sigma(\tau), t_k)} \right. \right.$$

$$\left. \left. + \overline{F}^\alpha(\tau) \frac{\cosh_r(\sigma(\tau), t_k)}{E_r(\sigma(\tau), t_k)} \right) \Delta_H \tau \right\}.$$

Thus, for $\alpha \in [0, 1]$, we obtain the (i)-solution of LIFDE (5.5) on $\mathbb{J}_k$ for $k \in \mathbb{N}_0$, when $r(t) < 0$ with $t \in \mathbb{T}^+$ as follows:

$$U(t) = U_k(t) = \cosh_r(t, t_k^+) \left\{ L_k U_k(t_k) + \int_{t_k}^t F(\tau) \left(\frac{\cosh_r(\sigma(\tau), t_k)}{E_r(\sigma(\tau), t_k)} \right. \right.$$

$$\left. \left. -_H \frac{\sinh_r(\sigma(\tau), t_k)}{E_r(\sigma(\tau), t_k)} \right) \Delta_H \tau \right\}$$

$$+ \sinh_r(t, t_k^+) \left\{ L_k U_k(t_k) + \int_{t_k}^t F(\tau) \left(\frac{\cosh_r(\sigma(\tau), t_k)}{E_r(\sigma(\tau), t_k)} \right. \right.$$

$$\left. \left. -_H \frac{\sinh_r(\sigma(\tau), t_k)}{E_r(\sigma(\tau), t_k)} \right) \Delta_H \tau \right\}.$$

Similarly, if $t \in \mathbb{J}^-$, we have

$$U(t) = V(t) = \cosh_r(t, 0)$$

$$\times \left\{ U_0 + \int_0^t F(\tau) \left(\frac{\cosh_r(\sigma(\tau), 0)}{E_r(\sigma(\tau), 0)} -_H \frac{\sinh_r(\sigma(\tau), 0)}{E_r(\sigma(\tau), 0)} \right) \Delta_H \tau \right\}$$

$$+ \sinh_r(t, 0)$$

$$\times \left\{ U_0 + \int_0^t F(\tau) \left(\frac{\cosh_r(\sigma(\tau), 0)}{E_r(\sigma(\tau), 0)} -_H \frac{\sinh_r(\sigma(\tau), 0)}{E_r(\sigma(\tau), 0)} \right) \Delta_H \tau \right\}.$$

Note that the H-difference

$$F(\tau) \frac{\cosh_r(\sigma(\tau), s)}{E_r(\sigma(\tau), s)} -_H F(\tau) \frac{\sinh_r(\sigma(\tau), s)}{E_r(\sigma(\tau), s)}$$

exists for $s \in [0, \sigma(\tau)]$ and $r(t) < 0$. The diameters of the α-level sets of

$$F(\tau) \frac{\cosh_r(\sigma(\tau), s)}{E_r(\sigma(\tau), s)} \quad \text{and} \quad F(\tau) \frac{\sinh_r(\sigma(\tau), s)}{E_r(\sigma(\tau), s)}$$

are, respectively,

$$\text{diam}\,([F(\tau)]^\alpha) \left[\frac{\cosh_r(\sigma(\tau), s)}{E_r(\sigma(\tau), s)} \right] \quad \text{and}$$

$$\text{diam}\,([F(\tau)]^\alpha) \left[\frac{-\sinh_r(\sigma(\tau), s)}{E_r(\sigma(\tau), s)} \right].$$

Since

$$\frac{\sinh_r(\sigma(\tau), s)}{E_r(\sigma(\tau), s)} + \frac{\cosh_r(\sigma(\tau), s)}{E_r(\sigma(\tau), s)} = \frac{e_r(\sigma(\tau), s)}{E_r(\sigma(\tau), s)} = \frac{1}{e_{-r}(\sigma(\tau), s)} > 0$$

for $r \in \mathcal{R}_1^+$ with $r(t) < 0$, we find that

$$\text{diam}([F(\tau)]^\alpha) \left[\frac{\cosh_r(\sigma(\tau), s)}{E_r(\sigma(\tau), s)} \right] > \text{diam}([F(\tau)]^\alpha) \left[\frac{-\sinh_r(\sigma(\tau), s)}{E_r(\sigma(\tau), s)} \right].$$

Again, since

$$\cosh_r(\sigma(t), s) \cosh_r^{\Delta t}(t, s) = r \cosh_r(\sigma(t), s) \sinh_r(t, s) > 0$$

and

$$\sinh_r(\sigma(t), s) \sinh_r^{\Delta t}(t, s) = r(t) \sinh_r(\sigma(t), s) \cosh_r(t, s) > 0$$

for $r(t) < 0$, in view of Lemma 5.1 (a), we see that $U(t)$ is (i)-differentiable on $\mathbb{T}^+ \setminus \mathbb{J}$. Consequently, LIFDE (5.5) has a (i)-solution and (5.10) holds.

(ii) For $t \in \mathbb{J}_k$, under the hypothesis of (ii)-differentiability, LIFDE (5.5) is transformed into the system of dynamic equations corresponding to linear impulsive dynamic equations (5.6) with A and f given as (5.18) and (5.19), respectively. Obviously, $A \in \mathcal{R}_n$ and

$$e_A(t, s) = e_r(t, s)\mathcal{I}.$$

By means of (5.7), we have

$$\begin{cases} u_k(t) = e_r(t, t_k^+) \left(\mathcal{I}\Phi_k + \int_{t_k}^t e_{\ominus r}(\sigma(\tau), t_k)\mathcal{I}f(\tau)\Delta_H\tau \right) & \text{for } t \in \mathbb{J}_k \\ u_k(t_k) = u_{k-1}(t_k) & \text{for } t_k \in \mathbb{T}^+. \end{cases}$$

Repeating the computations done for (i), we obtain the solution of corresponding system of linear dynamic equations, which is given by

$$[U_k(t)]^\alpha = \left[\underline{U_k}^\alpha(t), \overline{U_k}^\alpha(t) \right],$$

where

$$\underline{U_k}^\alpha(t) = e_r(t, t_k^+) \left(L_k \underline{U_k}^\alpha(t_k) + \int_{t_k}^t \overline{F}^\alpha(\tau)e_{\ominus r}(\sigma(\tau), t_k)\Delta_H\tau \right)$$

and

$$\overline{U_k}^\alpha(t) = e_r(t, t_k^+) \left(L_k \overline{U_k}^\alpha(t_k) + \int_{t_k}^t \underline{F}^\alpha(\tau)e_{\ominus r}(\sigma(\tau), t_k)\Delta_H\tau \right)$$

for $\alpha \in [0, 1]$. We assert that the (ii)-solution of LIFDE (5.5) on $\mathbb{J}_k$ is

$$U(t) = U_k(t) = e_r(t, t_k^+) \left(L_k U_k(t_k) -_H \int_{t_k}^t -F(\tau)e_{\ominus r}(\sigma(\tau), t_k)\Delta_H\tau \right),$$

where $k \in \mathbb{N}_0$ and $r(t) < 0$. In fact, we observe that

$$e_r(\sigma(t), t_k)e_r^{\Delta t}(t, t_k) = r(t)e_r(\sigma(t), t_k)e_r(t, t_k) < 0.$$

If we denote

$$G_k(t) = L_k U_k(t_k) -_H \int_{t_k}^t -F(\tau)e_{\ominus r}(\sigma(\tau), t_k)\Delta_H\tau,$$

then Theorem 5.3(i) guarantees that G_k is (ii)-differentiable and

$$\Delta_H G_k(t) = F(t)e_{\ominus r}(\sigma(t), t_k).$$

Now, the conditions in Lemma 5.1(b) are satisfied. Thus,

$$\begin{aligned} \Delta_H U_k(t) &= r(t)e_r(t, t_k^+)G_k(t) + e_r(\sigma(t), t_k)\Delta_H G_k(t) \\ &= r(t)U_k(t) + F(t) \end{aligned}$$

for $k \in \mathbb{N}_0$. Similarly, for $t \in \mathbb{J}^-$, we have

$$U(t) = V(t) = e_r(t,0) \left(U_0 -_H \int_0^t -F(\tau) e_{\ominus r}(\sigma(\tau),0) \Delta_H \tau \right).$$

Thus, we obtain that LIFDE (5.5) has the (ii)-solution satisfying (5.11). This completes the proof. □

Theorem 5.5. *Suppose that* $-r \in \mathcal{R}_1^+$ *satisfies* $r(t) > 0$ *for all* $t \in \mathbb{T}^+$. *Then,*

(i) *the LIFDE (5.5) has the (ii)-solution on* $\mathbb{T}^+$ *given as in (5.10), where*

$$V(t) = \cosh_r(t,0) \left\{ U_0 -_H \int_0^t \left(F(\tau) \frac{\sinh_r(\sigma(\tau),0)}{E_r(\sigma(\tau),0)} \right. \right.$$

$$\left. -_H F(\tau) \frac{\cosh_r(\sigma(\tau),0)}{E_r(\sigma(\tau),0)} \right) \Delta_H \tau \Big\}$$

$$-_H \left[-\sinh_r(t,0) \right] \left\{ U_0 -_H \int_0^t \left(F(\tau) \frac{\sinh_r(\sigma(\tau),0)}{E_r(\sigma(\tau),0)} \right. \right.$$

$$\left. -_H F(\tau) \frac{\cosh_r(\sigma(\tau),0)}{E_r(\sigma(\tau),0)} \right) \Delta_H \tau \Big\},$$

and

$$U_k(t) = \cosh_r(t,t_k^+) \left\{ L_k U_k(t_k) -_H \int_{t_k}^t \left(F(\tau) \frac{\sinh_r(\sigma(\tau),t_k)}{E_r(\sigma(\tau),t_k)} \right. \right.$$

$$\left. -_H F(\tau) \frac{\cosh_r(\sigma(\tau),t_k)}{E_r(\sigma(\tau),t_k)} \right) \Delta_H \tau \Big\} -_H \left[-\sinh_r(t,t_k^+) \right]$$

$$\times \left\{ L_k U_k(t_k) -_H \int_{t_k}^t \left(F(\tau) \frac{\sinh_r(\sigma(\tau),t_k)}{E_r(\sigma(\tau),t_k)} \right. \right.$$

$$\left. -_H F(\tau) \frac{\cosh_r(\sigma(\tau),t_k)}{E_r(\sigma(\tau),t_k)} \right) \Delta_H \tau \Big\},$$

provided that the H-differences exist and

$$U_0(t_0) = V(t_0), \quad U_k(t_k) = U_{k-1}(t_k) \quad for\ k \in \mathbb{N}_0.$$

(ii) *the LIFDE* (5.5) *has the* (i)*-solution on* $\mathbb{T}^+$ *is given by* (5.11), *where*

$$V(t) = e_r(t,0)\left(U_0 + \int_0^t F(\tau)e_{\ominus r}(\sigma(\tau),0)\Delta_H\tau\right)$$

and

$$U_k(t) = e_r(t,t_k^+)\left(L_k U_k(t_k) + \int_{t_k}^t F(\tau)e_{\ominus r}(\sigma(\tau),t_k)\Delta_H\tau\right)$$

where $U_0(t_0) = V(t_0)$, $U_k(t_k) = U_{k-1}(t_k)$ *for* $k \in \mathbb{N}_0$.

Proof. (i) The LIFDE(5.5) with (ii)-differentiability is transformed into linear impulsive dynamic equation system (5.8) with $u(t)$, $u_{t_k^+}$, $A(t)$, and v_0 given in (5.13), (5.14), (5.17), respectively, and $f(t)$ as in (5.19). By (5.8), we obtain

$$\begin{cases} u_k(t) = e_A(t,t_k^+)\left(\Phi_k + \int_{t_k}^t e_A(t_k,\sigma(\tau))f(\tau)\Delta_H\tau\right) & \text{for } t \in \mathbb{J}_k \\ u_k(t_k) = u_{k-1}(t_k) & \text{for } t_k \in \mathbb{T}^+, \end{cases}$$

where $u_{0-1}(t_0) = v(t_0)$. Repeating the computations done for (i) of Theorem 5.4, we obtain the solution of corresponding system of linear dynamic equations, which is given by

$$[U(t)]^\alpha = \left[\underline{U}^\alpha(t), \overline{U}^\alpha(t)\right],$$

where

$$\underline{U}^\alpha(t) = \underline{U_k}^\alpha(t)$$

$$= \cosh_r(t,t_k^+)\left\{L_k\underline{U_k}^\alpha(t_k) + \int_{t_k}^t \left(\overline{F}^\alpha(\tau)\frac{\cosh_r(\sigma(\tau),t_k)}{E_r(\sigma(\tau),t_k)}\right.\right.$$

$$\left.-\underline{F}^\alpha(\tau)\frac{\sinh_r(\sigma(\tau),t_k)}{E_r(\sigma(\tau),t_k)}\right)\Delta_H\tau\Big\}$$

$$+ \sinh_r(t,t_k^+)\left\{L_k\overline{U_k}^\alpha(t_k) + \int_{t_k}^t \left(-\overline{F}^\alpha(\tau)\frac{\sinh_r(\sigma(\tau),t_k)}{E_r(\sigma(\tau),t_k)}\right.\right.$$

$$\left.+\underline{F}^\alpha(\tau)\frac{\cosh_r(\sigma(\tau),t_k)}{E_r(\sigma(\tau),t_k)}\right)\Delta_H\tau\Big\}$$

and

$$\overline{U}^\alpha(t) = \overline{U_k}^\alpha(t)$$

$$= \sinh_r(t, t_k^+) \left\{ L_k \underline{U_k}^\alpha(t_k) + \int_{t_k}^t \left(\overline{F}^\alpha(\tau) \frac{\cosh_r(\sigma(\tau), t_k)}{E_r(\sigma(\tau), t_k)} \right. \right.$$

$$\left. - \underline{F}^\alpha(\tau) \frac{\sinh_r(\sigma(\tau), t_k)}{E_r(\sigma(\tau), t_k)} \right) \Delta_H \tau \right\}$$

$$+ \cosh_r(t, t_k^+) \left\{ L_k \overline{U_k}^\alpha(t_k) + \int_{t_k}^t \left(-\overline{F}^\alpha(\tau) \frac{\sinh_r(\sigma(\tau), t_k)}{E_r(\sigma(\tau), t_k)} \right. \right.$$

$$\left. + \underline{F}^\alpha(\tau) \frac{\cosh_r(\sigma(\tau), t_k)}{E_r(\sigma(\tau), t_k)} \right) \Delta_H \tau \right\}.$$

for $\alpha \in [0, 1]$, Next, for $\alpha \in [0, 1]$, we show that the (ii)-solution of LIFDE (5.5) on $\mathbb{J}_k$ for $k \in \mathbb{N}_0$, when $r(t) > 0$ with $t \in \mathbb{T}^+$, will be given as follows:

$$U_k(t) = \cosh_r(t, t_k^+) G_k(t) -_H (-\sinh_r(t, t_k^+)) G_k(t),$$

where

$$G_k(t) = L_k U_k(t_k) -_H$$

$$\times \int_{t_k}^t F(\tau) \left(\frac{\sinh_r(\sigma(\tau), t_k)}{E_r(\sigma(\tau), t_k)} -_H \frac{\cosh_r(\sigma(\tau), t_k)}{E_r(\sigma(\tau), t_k)} \right) \Delta_H \tau$$

is (ii)-differentiable on $\mathbb{J}_k$ for $k \in \mathbb{N}_0$.

First, by our hypothesis G_k is well-defined. Second, Theorem 5.3 (i) guarantees that G_k is (ii)-differentiable and $\text{diam}[G_k(t)]$ is nonincreasing in t for fixed $\alpha \in [0, 1]$. Note that

$$\cosh_r(t, t_k^+) - \sinh_r(t, t_k^+) = e_{-r}(t, t_k^+)$$

is nonnegative and decreasing in t. Thus, $\text{diam}[U_k(t)]^\alpha$ is nonincreasing in t for fixed $\alpha \in [0, 1]$. Hence, the H-differences

$$U_k(\sigma(t)) -_H U_k(t + h) \quad \text{and} \quad U_k(t - h) -_H U_k(\sigma(t))$$

exist. Third, we check that the (ii)-derivative of U_k is $r(t)U_k(t) + F(t)$ for $t \in \mathbb{J}_k$, $k \in \mathbb{N}_0$. If t is right-dense, then, we find that

$$\lim_{h \to 0^+} \frac{U_k(t) -_H U_k(t + h)}{-h} = \lim_{h \to 0^+} \frac{U_k(t - h) -_H U_k(t)}{-h}$$

$$= r(t)U_k(t) + F(t).$$

In the view of Proposition 2.3(III), we have

$$\Delta_H U_k(t) = r(t)U_k(t) + F(t).$$

If t is right-scattered, then we write

$$\xi_k(t) = \int_{t_k}^{t} \left(\overline{F}^\alpha(\tau)\frac{\cosh_r(\sigma(\tau), t_k)}{E_r(\sigma(\tau), t_k)} - \underline{F}^\alpha(\tau)\frac{\sinh_r(\sigma(\tau), t_k)}{E_r(\sigma(\tau), t_k)} \right) \Delta_H \tau$$

and

$$\eta_k(t) = \int_{t_k}^{t} \left(-\overline{F}^\alpha(\tau)\frac{\sinh_r(\sigma(\tau), t_k)}{E_r(\sigma(\tau), t_k)} + \underline{F}^\alpha(\tau)\frac{\cosh_r(\sigma(\tau), t_k)}{E_r(\sigma(\tau), t_k)} \right) \Delta_H \tau.$$

This gives

$$\xi_k^\Delta(t) = \overline{F}^\alpha(t)\frac{\cosh_r(\sigma(t), t_k)}{E_r(\sigma(t), t_k)} - \underline{F}^\alpha(t)\frac{\sinh_r(\sigma(t), t_k)}{E_r(\sigma(t), t_k)}$$

and

$$\eta_k^\Delta(t) = -\overline{F}^\alpha(t)\frac{\sinh_r(\sigma(t), t_k)}{E_r(\sigma(t), t_k)} + \underline{F}^\alpha(t)\frac{\cosh_r(\sigma(t), t_k)}{E_r(\sigma(t), t_k)}$$

Hence,

$$\xi_k(\sigma(t)) - \xi_k(t)$$

$$= \int_{t}^{\sigma(t)} \left(\overline{F}^\alpha(\tau)\frac{\cosh_r(\sigma(\tau), t_k)}{E_r(\sigma(\tau), t_k)} - \underline{F}^\alpha(\tau)\frac{\sinh_r(\sigma(\tau), t_k)}{E_r(\sigma(\tau), t_k)} \right) \Delta_H \tau$$

$$= \mu(t)\overline{F}^\alpha(t)\frac{\cosh_r(\sigma(t), t_k)}{E_r(\sigma(t), t_k)} - \mu(t)\underline{F}^\alpha(t)\frac{\sinh_r(\sigma(t), t_k)}{E_r(\sigma(t), t_k)},$$

and

$$\eta_k(\sigma(t)) - \eta_k(t)$$

$$= -\mu(t)\overline{F}^\alpha(t)\frac{\sinh_r(\sigma(t), t_k)}{E_r(\sigma(t), t_k)} + \mu(t)\underline{F}^\alpha(t)\frac{\cosh_r(\sigma(t), t_k)}{E_r(\sigma(t), t_k)}.$$

Next, from Proposition 2.3(II) and the property (p2), it follows that

$$\sinh_r(\sigma(t), t_k) - \sinh_r(t+h, t_k) = r(t)\cosh_r(t, t_k)\mu(t)$$

and

$$\cosh_r(\sigma(t), t_k) - \cosh_r(t+h, t_k) = r(t)\sinh_r(t, t_k)\mu(t).$$

Due to the above arguments, for each $\alpha \in [0,1]$, we uniformly have

$$\underline{U_k}^\alpha(\sigma(t)) - \underline{U_k}^\alpha(t) - \left(r(t)\overline{U_k}^\alpha(t) + \overline{F}^\alpha(t)\right)\mu(t)$$

$$= \cosh_r(\sigma(t), t_k^+)\left(L_k\underline{U_k}^\alpha(t_k) + \xi_k(\sigma(t))\right)$$

$$+ \sinh_r(\sigma(t), t_k^+)\left(L_k\overline{U_k}^\alpha(t_k) + \eta_k(\sigma(t))\right)$$

$$- \cosh_r(t, t_k^+)\left(L_k\underline{U_k}^\alpha(t_k) + \xi_k(t)\right)$$

$$- \sinh_r(t, t_k^+)\left(L_k\overline{U_k}^\alpha(t_k) + \eta_k(t)\right)$$

$$- r(t)\mu(t)\left\{\sinh_r(t, t_k^+)\left(L_k\underline{U_k}^\alpha(t_k) + \xi_k(t)\right)\right.$$

$$\left. + \cosh_r(t, t_k^+)\left(L_k\overline{U_k}^\alpha(t_k) + \eta_k(t)\right)\right\} - \overline{F}^\alpha(t)\mu(t)$$

$$= \left(\cosh_r(\sigma(t), t_k) - \cosh_r(t, t_k) - r(t)\mu(t)\sinh_r(t, t_k)\right)$$

$$\times \left(L_k\underline{U_k}^\alpha(t_k) + \xi_k(\sigma(t))\right)$$

$$+ \left(\sinh_r(\sigma(t), t_k) - \sinh_r(t, t_k) - r(t)\mu(t)\cosh_r(t, t_k)\right)$$

$$\times \left(L_k\overline{U_k}^\alpha(t_k) + \eta_k(\sigma(t))\right) + \cosh_r(t, t_k)(\xi_k(\sigma(t)) - \xi_k(t))$$

$$+ r(t)\mu(t)\sinh_r(t, t_k)\left(\xi_k(\sigma(t)) - \xi_k(t)\right)$$

$$+ \sinh_r(t, t_k)(\eta_k(\sigma(t)) - \eta_k(t))$$

$$+ r(t)\mu(t)\cosh_r(t, t_k)\left(\eta_k(\sigma(t)) - \eta_k(t)\right) - \overline{F}^\alpha(t)\mu(t).$$

$$= \mu(t)\left(\cosh_r(t, t_k) + r(t)\sinh_r(t, t_k)\right)$$

$$\times \left(\overline{F}^\alpha(t)\frac{\cosh_r(\sigma(t), t_k)}{E_r(\sigma(t), t_k)} - \underline{F}^\alpha(t)\frac{\sinh_r(\sigma(t), t_k)}{E_r(\sigma(t), t_k)}\right)$$

$$+ \mu(t)\left(\sinh_r(t, t_k) + r(t)\cosh_r(t, t_k)\right)$$

$$\times \left(-\overline{F}^\alpha(t)\frac{\sinh_r(\sigma(t), t_k)}{E_r(\sigma(t), t_k)} + \underline{F}^\alpha(t)\frac{\cosh_r(\sigma(t), t_k)}{E_r(\sigma(t), t_k)}\right) - \overline{F}^\alpha(t)\mu(t)$$

$$= \mu(t)\cosh_r(\sigma(t), t_k)$$

$$\times \left(\overline{F}^\alpha(t)\frac{\cosh_r(\sigma(t), t_k)}{E_r(\sigma(t), t_k)} - \underline{F}^\alpha(t)\frac{\sinh_r(\sigma(t), t_k)}{E_r(\sigma(t), t_k)}\right)$$

$$+ \mu(t)\sinh_r(\sigma(t), t_k)$$

$$\times \left(-\overline{F}^\alpha(t)\frac{\sinh_r(\sigma(t), t_k)}{E_r(\sigma(t), t_k)} + \underline{F}^\alpha(t)\frac{\cosh_r(\sigma(t), t_k)}{E_r(\sigma(t), t_k)}\right) - \overline{F}^\alpha(t)\mu(t)$$

$$= 0.$$

That is,

$$\underline{U_k}^\alpha(\sigma(t)) - \underline{U_k}^\alpha(t) = \left(r(t)\overline{U_k}^\alpha(t) + \overline{F}^\alpha(t)\right)\mu(t).$$

Analogously, we can prove

$$\overline{U_k}^\alpha(\sigma(t)) - \overline{U_k}^\alpha(t) = \left(r(t)\underline{U_k}^\alpha(t) + \underline{F}^\alpha(t)\right)\mu(t)$$

for all $\alpha \in [0,1]$. Hence, Proposition 2.3(II) guarantees that

$$\Delta_H U_k(t) = r(t)U_k(t) + F(t)$$

for all right-scattered point $t \in \mathbb{J}_k$. Consequently, U_k is a (ii)-solution of LIFDE (5.5) on $\mathbb{J}_k$ for $k \in \mathbb{N}_0$. Similarly, for $t \in \mathbb{J}^-$, we have

$$
\begin{aligned}
U(t) = \cosh_r(t) &\left\{ U_0 -_H \int_0^t F(\tau) \left(\frac{\sinh_r(\sigma(\tau), 0)}{E_r(\sigma(\tau), 0)} \right. \right. \\
&\left. \left. -_H \frac{\cosh_r(\sigma(\tau), 0)}{E_r(\sigma(\tau), 0)} \right) \Delta\tau \right\} \\
-_H[- \sinh_r(t)] &\left\{ U_0 -_H \int_0^t F(\tau) \left(\frac{\sinh_r(\sigma(\tau), 0)}{E_r(\sigma(\tau), 0)} \right. \right. \\
&\left. \left. -_H \frac{\cosh_r(\sigma(\tau), 0)}{E_r(\sigma(\tau), 0)} \right) \Delta\tau \right\} \\
= V(t). &
\end{aligned}
\tag{5.20}
$$

Now, both equations in (i) hold on $\mathbb{T}^+$ and hence (i) is proved.

(ii) For $t \in \mathbb{J}_k$, under the hypothesis of (i)-differentiability, the LIFDE (5.5) is transformed into the linear impulsive dynamic system corresponding to (5.6) with $A \in \mathcal{R}_n$ given in (5.18) and f given in (5.15). Substituting $e_A(t, s) = e_r(t, s)\mathcal{I}$ in (5.7) and repeating the process of the proof of Theorem 5.4, we obtain, for all $\alpha \in [0,1]$,

$$\underline{U_k}^\alpha(t) = e_r(t, t_k^+) \left(L_k \underline{U_k}^\alpha(t_k) - \int_{t_k}^t -\overline{F}^\alpha(\tau)e_{\ominus r}(\sigma(\tau), t_k)\Delta_H\tau \right)$$

and

$$\overline{U_k}^\alpha(t) = e_r(t, t_k^+) \left(L_k \overline{U_k}^\alpha(t_k) - \int_{t_k}^t -\underline{F}^\alpha(\tau)e_{\ominus r}(\sigma(\tau), t_k)\Delta_H\tau \right).$$

We now show that the (i)-solution of LIFDE (5.5) on J_k for $k \in \mathbb{N}_0$, when $r(t) > 0$ with $t \in \mathbb{T}^+$ will be given as follows:

$$U_k(t) = e_r(t, t_k^+) \left(L_k U_k(t_k) + \int_{t_k}^t F(\tau) e_{\ominus r}(\sigma(\tau), t_k) \Delta_H \tau \right).$$

Note that

$$e_r(\sigma(t), t_k) e_r^\Delta(t, t_k) = r(t) e_r(\sigma(t), t_k) e_r(t, t_k) > 0$$

for $r(t) > 0$. From Theorem 5.3(v), it follows that

$$L_k U_k(t_k) + \int_{t_k}^t F(\tau) e_{\ominus r}(\sigma(\tau), t_k) \Delta_H \tau$$

is (i)-differentiable on $\mathbb{T}^+$ and from Lemma 5.1(a), it follows that

$$\Delta_H U_k(t) = r(t) e_r(t, t_k^+) \left(L_k U_k(t_k) + \int_{t_k}^t F(\tau) e_{\ominus r}(\sigma(\tau), t_k) \Delta_H \tau \right)$$

$$+ e_r(\sigma(t), t_k^+) F(t) e_{\ominus r}(\sigma(t), t_k)$$

$$= r(t) U_k(t) + F(t)$$

for $k \in \mathbb{N}_0$. Similarly, we have for $t \in \mathbb{J}^-$

$$U(t) = V(t) = e_{\sim}(t) \left(U_0 + \int_0^t F(\tau) e_{\ominus r}(\sigma(\tau)) \Delta \tau \right).$$

This shows that both equations in (ii) hold on $\mathbb{T}^+$ and hence (ii) is proved. This completes the proof. □

From Theorems 5.4 and 5.5, the proof of the following theorem follows easily.

Theorem 5.6. *Assume that* $r \equiv 0$. *Then,*

(i) *the LIFDE* (5.5) *has the* (i)-*solution given by*

$$U(t) = \begin{cases} V(t) & for\ t \in \mathbb{J}^- \\ U_k(t) & for\ t \in \mathbb{J}_k,\ k \in \mathbb{N}_0, \end{cases}$$

where

$$V(t) = U_0 + \int_0^t F(\tau) \Delta_H \tau$$

and

$$U_k(t) = L_k U_k(t_k) + \int_{t_k}^t F(\tau) \Delta_H \tau.$$

(ii) *the LIFDE* (5.5) *has the* (ii)-*solution given by*

$$U(t) = \begin{cases} V(t) & for\ t \in \mathbb{J}^- \\ U_k(t) & for\ t \in \mathbb{J}_k,\ k \in \mathbb{N}_0, \end{cases}$$

where

$$V(t) = U_0 -_H \int_0^t (-F(\tau))\Delta_H \tau$$

and

$$U_k(t) = L_k U_k(t_k) -_H \int_{t_k}^t (-F(\tau))\Delta_H \tau$$

provided the above H-differences exist and

$$U_0(t_0) = V(t_0), \quad U_k(t_k) = U_{k-1}(t_k) \quad for\ k \in \mathbb{N}_0.$$

Remark 5.2. If $\mathbb{T} = \mathbb{R}$, then

$$E_r(t, s) \equiv 1,$$

$$\sinh_r(t, s) = \sinh\left\{ \int_s^t r(\tau)d\tau \right\},$$

and

$$\cosh_r(t, s) = \cosh\left\{ \int_s^t r(\tau)d\tau \right\}.$$

Thus, our results are the extension and improvement of the corresponding results for fuzzy differential equations proposed by Khastan *et al.* (2011). In particular, the present results are identical to those in Khastan *et al.* (2011) in case that we consider t restricted to $\mathbb{J}^-$ and under H-difference.

If $\mathbb{T} = \mathbb{Z}$, then

$$E_r(t, s) = \prod_{\tau=s}^{t-1} \left(1 - r^2(\tau)\right),$$

$$\sinh_r(t, s) = \frac{1}{2} \left(\prod_{\tau=s}^{t-1} (1 + r(\tau)) - \prod_{\tau=s}^{t-1} (1 - r(\tau)) \right),$$

and

$$\cosh_r(t, s) = \frac{1}{2} \left(\prod_{\tau=s}^{t-1} (1 + r(\tau)) + \prod_{\tau=s}^{t-1} (1 - r(\tau)) \right)$$

provided that $r \neq \pm 1$ and $s < t$. Thus, we obtain the general form of solutions to linear impulsive fuzzy difference equation

$$\begin{cases} U(n+1) = r(n)U(n) + U(n) + F(n) & \text{for } n \in \mathbb{Z}^+ \setminus \mathbb{J} \\ U_{t_k^+} = L_k U(t_k) & \text{for } t_k \in \mathbb{J},\ k \in \mathbb{N}_0 \\ U(0) = U_0 \in \mathbb{T}_f & \text{for } t_0 \in \mathbb{T}^+. \end{cases}$$

Moreover, we also extend these "classical cases" to the cases "in between". For instance, take

$$\mathbb{T} = \left\{ \sum_{k=1}^{n} \frac{1}{k} : n \in \mathbb{N} \right\}.$$

In this case,

$$E_r(t, s) = \binom{n - s + r}{n - s} \binom{n - s - r}{n - s}$$

and

$$\sinh_r(t, s) = \frac{1}{2} \left[\binom{n - s + r}{n - s} - \binom{n - s - r}{n - s} \right],$$

and

$$\cosh_r(t, s) = \frac{1}{2} \left[\binom{n - s + r}{n - s} + \binom{n - s - r}{n - s} \right]$$

where $t = \sum_{k=1}^{n} \frac{1}{k}$. We can also consider the so-called q-difference problems.

Remark 5.3. Although the linear impulsive dynamic problem (5.8) has a unique solution on $\mathbb{T}^+$, the solution of LIFDE (5.5) is not unique in general.

5.3   Illustrative Examples

In this section, we present couple of examples to further illustrate the applicability of the results involved in the previous sections.

Example 5.1. Consider the impulsive fuzzy dynamic equation

$$\begin{cases} \Delta_H U(t) = aU(t) + t & \text{for } t \in \mathbb{T}^+ \setminus \mathbb{J} \\ U(t_k^+) = L_k U(t_k) & \text{for } t_k \in \mathbb{J}, \ k \in \mathbb{N}_0 \\ U(t_0) = U_0, \end{cases} \tag{5.21}$$

where $a < 0$, $a \in \mathcal{R}_+^1$, and L_k is bounded linear operator. It is easy to infer the existence of H-differences

$$U_0 -_H \int_0^t -\tau e_{\ominus a}(\sigma(\tau)) \Delta \tau$$

and

$$L_k U(t_k) -_H \int_{t_k}^t -\tau e_{\ominus a}(\sigma(\tau), t_k) \Delta \tau.$$

Thus, by Theorem 5.4, the (ii)-solution of LIFDE (5.21) on $\mathbb{T}^+$ is given by

$$U(t) = \begin{cases} V(t) & \text{for } t \in \mathbb{J}^- \\ U_k(t) & \text{for } t \in \mathbb{J}_k, \end{cases} \tag{5.22}$$

where

$$V(t) = e_a(t) \left[U_0 -_H \int_0^t -\tau e_{\ominus a}(\sigma(\tau)) \Delta \tau \right]$$

and

$$U_k(t) = e_a(t, t_k^+) \left[L_k U_k(t_k) -_H \int_{t_k}^t -\tau e_{\ominus a}(\sigma(\tau), t_k) \Delta \tau \right],$$

$U_0(t_0) = V(t_0)$, $U_k(t_k) = U_{k-1}(t_k)$ for $k \in \mathbb{N}_0$.

Remark 5.4. If $\mathbb{T} = \mathbb{R}$, then we get

$$V(t) = e^{at} \left[U_0 -_H \int_0^t -\tau e^{-a\tau} d\tau \right] \quad \text{for } t \in \mathbb{J}^-,$$

and

$$U_k(t) = e^{a(t-t_k)} \left[L_k U_k(t_k) -_H \int_{t_k}^t -\tau e^{-a(\tau-t_k)} d\tau \right] \quad \text{for } t \in \mathbb{J}_k, \ k \in \mathbb{N}_0.$$

By performing simple computations, for $t \in \mathbb{J}_k$, we obtain

$$U_k(t) = \frac{1}{a}\left(t + \frac{1}{a}\right)\widetilde{1} + \left[L_k U_k(t_k) + \left(\frac{t}{a} + \frac{1}{a^2}\right) \right] e^{a(t-t_k)}$$

with $\widetilde{1} = \chi_{\{1\}}$. In this case, we observe that

$$D\left[U(t), \frac{1}{a}\left(t + \frac{1}{a}\right)\widetilde{1}\right] \le \left\| L_k U_k(t_k) + \frac{t}{a} + \frac{1}{a^2} \right\| e^{at}$$

and this implies that

$$\lim_{t\to\infty} D\left[U(t), \frac{1}{a}\left(t + \frac{1}{a}\right)\widetilde{1}\right] = 0.$$

Thus, in the continuous case, the uncertainty asymptotically disappears on the fuzzy system.

If $\mathbb{T} = \mathbb{Z}$, then we get

$$V(t) = (1+a)^t \left[U_0 -_H \sum_{\tau=0}^{t-1} (-\tau)(1+a)^{-(\tau+1)} \right] \quad \text{for } t \in \mathbb{J}^-,$$

and

$$U_k(t) = (1+a)^{(t-t_k)} \left[L_k U_k(t_k) -_H \sum_{\tau=t_k}^{t-1} (-\tau)(1+a)^{-(\tau+1-t_k)} \right]$$

$$\text{for } t \in \mathbb{J}_k, \ k \in \mathbb{N}_0$$

By performing simple computations, for $t \in \mathbb{J}_k$, we obtain

$$U(t) = U_k(t) = \frac{(1+a)^{1-t_k}}{a}\widetilde{1} + \left[L_k U(t_k) - \frac{1+a}{a} \right] (1+a)^{t-t_k}$$

with $\widetilde{1} = \chi_{\{1\}}$. Therefore, in the discrete case, the phenomenon that the uncertainty asymptotically disappears on the fuzzy system arises only $a > -1$.

Example 5.2. Consider the LIFDE

$$\begin{cases} \Delta_H U(t) = 2tU(t) + t\gamma & \text{for } t \in \mathbb{T}^+ \setminus \mathbb{J} \\ U_{t_k^+} = 4U(t_k) & \text{for } t_k \in \mathbb{J}, \ k \in \mathbb{N}_0 \\ U(0) = \gamma, \end{cases} \qquad (5.23)$$

where $[\gamma]^\alpha = [\alpha - 1, 1 - \alpha]$ with $\alpha \in [0, 1]$.

Let $\mathbb{T} = \mathbb{R}$ and $t_k = \sqrt{(k+1)\ln(2)}$ for $k \in \mathbb{N}_0$. As a result of Theorem 5.5, the (i)-solution of (5.23) is given by

$$U(t) = \begin{cases} V(t) & \text{for } t \in \mathbb{J}^- = [0, \sqrt{\ln 2}] \cap \mathbb{T}^+ \\ U_k(t) & \text{for } t \in \mathbb{J}_k, \ k \in \mathbb{N}_0, \end{cases}$$

where

$$V(t) = \frac{1}{2}(3e^{t^2} - 1)\gamma$$

and

$$U_k(t) = 4U_k\left(\sqrt{(k+1)\ln 2}\right)e^{t^2 - (k+1)\ln 2} + \frac{1}{2}\left(e^{t^2 - (k+1)\ln 2} - 1\right)\gamma.$$

To seek the (ii)-solution, we note that

$$\cosh_r(t, t_k) = \cosh(t^2 - t_k^2)$$

and

$$\sinh_r(t, t_k) = \sinh(t^2 - t_k^2),$$

when $r(t) = 2t$. The solution of the system of dynamic equations corresponding to (5.7) is given by

$$[U_k(t)]^\alpha = \left[\underline{U_k}^\alpha(t), \overline{U_k}^\alpha(t)\right],$$

where

$$\underline{U_k}^\alpha(t) = 4\underline{U_k}^\alpha(t_k)\cosh(t^2 - t_k^2) + 4\overline{U_k}^\alpha(t_k)\sinh(t^2 - t_k^2)$$
$$- \frac{\alpha - 1}{2}\left[\cosh(t^2 - t_k^2) + \sinh(t^2 - t_k^2)) - 1\right]$$
$$\times (\cosh(t^2 - t_k^2) - \sinh(t^2 - t_k^2)),$$

and

$$\overline{U_k}^\alpha(t) = 4\underline{U_k}^\alpha(t_k)\sinh(t^2 - t_k^2) + 4\overline{U_k}^\alpha(t_k)\cosh(t^2 - t_k^2)$$
$$- \frac{1-\alpha}{2}\left[\cosh(t^2 - t_k^2) + \sinh(t^2 - t_k^2)) - 1\right]$$
$$\times (\cosh(t^2 - t_k^2) - \sinh(t^2 - t_k^2)).$$

Note that the H-difference

$$\gamma -_H \int_0^t [\tau\gamma\sinh(\tau^2) - \tau\gamma\cosh(\tau^2)]d\tau$$

exists on $\left[0, \sqrt{\ln(2)}\right]$ and so the (ii)-solution $V(t)$ of (5.23) on $\left[0, \sqrt{\ln(2)}\right]$ can be written as

$$V(t) = \frac{1}{2}(3e^{-t^2} - 1)\gamma.$$

In particular, $V\left(\sqrt{\ln(2)}\right) = \frac{1}{4}\gamma$. From the fact $L_0 U_0(t_0) = 4 \times \frac{1}{4}\gamma = \gamma$, it follows that the H-difference

$$L_0 U_0(t_0) -_H \int_{t_0}^t [\tau\gamma\sinh(\tau^2 - t_0^2) - \tau\gamma\cosh(\tau^2 - t_0^2)]d\tau$$
$$= \gamma -_H \int_{\sqrt{\ln(2)}}^t [\tau\gamma\sinh(\tau^2 - \ln(2)) - \tau\gamma\cosh(\tau^2 - \ln(2))]d\tau$$

exists on $\left(\sqrt{\ln(2)}, \sqrt{2\ln(2)}\right]$ and the (ii)-solution $U_0(t)$ of (5.23) on $\left(\sqrt{\ln(2)}, \sqrt{2\ln(2)}\right]$ can be written as

$$U_0(t) = \cosh\left(t^2 - \ln(2)\right)\left[\gamma -_H \int_{\sqrt{\ln(2)}}^t [\tau\gamma\sinh\left(\tau^2 - \ln(2)\right)\right.$$
$$\left. -\tau\gamma\cosh\left(\tau^2 - \ln(2)\right)]d\tau\right] -_H \left(-\sinh\left(t^2 - \ln(2)\right)\right)$$
$$\times \left[\gamma -_H \int_{\sqrt{\ln 2}}^t [\tau\gamma\sinh\left(\tau^2 - \ln(2)\right)\right.$$

$$-\tau\gamma\cosh\left(\tau^2-\ln(2)\right)\Big]\,d\tau\Bigg]$$

$$=\left(1-\frac{1}{2}\left(\sinh\left(t^2-\ln(2)\right)+\cosh\left(t^2-\ln(2)\right)-1\right)\right)$$

$$\times\cosh\left(t^2-\ln(2)\right)-\sinh\left(t^2-\ln(2)\right)\gamma$$

$$=\frac{1}{2}\left(3e^{-(t^2-\ln(2))}-1\right).$$

In particular, $U_0\left(\sqrt{2\ln(2)}\right)=\frac{1}{4}\gamma$. On the analogy of this process, we obtain that the (ii)-solution of (5.23) on $\mathbb{J}_k=\left(\sqrt{(k+1)\ln(2)},\sqrt{(k+2)\ln(2)}\right]$ can be expressed by

$$U_k(t)=\frac{1}{2}\left(3e^{-(t^2-(k+1)\ln(2))}-1\right)\quad\text{for }t\in\mathbb{J}_k,$$

and

$$U_k\left(\sqrt{(k+2)\ln(2)}\right)=\frac{1}{4}\gamma\quad\text{for }k\in\mathbb{N}.$$

Let $\mathbb{T}=\mathbb{Z}$ and $t_k=2k+1$, $k\in\mathbb{N}_0$. Then, $r(t)=2t$ implies that

$$e_r(t,s)=\prod_{\tau=s}^{t-1}(1+2\tau).$$

Therefore, the (i)-solution of (5.23) is given by

$$U(t)=\begin{cases}V(t)&\text{for }t\in\mathbb{J}^-\\U_k(t)&\text{for }t\in\mathbb{J}_k,\ k\in\mathbb{N}_0.\end{cases}$$

where

$$V(t)=\gamma$$

and

$$U_k(t)=\prod_{\tau=2k+1}^{t-1}(1+2\tau)\times\left[4U_{k-1}(t_k)+\sum_{\tau=2k+1}^{t-1}\frac{\tau}{\prod_{s=2k+1}^{\tau}(1+2s)}\gamma\right].$$

Here, we have $\mathbb{J}^- = [0,1] \cap \mathbb{Z}^+$, $\mathbb{J}_k = (2k+1, 2k+3] \cap \mathbb{Z}^+$, $U_{0-1}(t_0) = \gamma$, and

$$U_0(t_1) = 3 \cdot 5(4\gamma) + \frac{1}{5}\gamma + 2\gamma, \dots.$$

Similarly, we can present the expression of the (ii)-solutions of (5.23) in the discrete case.

Appendix

In order to make this monograph self contained, here, we list some additional but important results that are used in this monograph.

There are two forms of Arzelà–Ascoli's theorem in context of time scales, and they are stated as follows:

Theorem A.1 (Arzelà–Ascoli theorem; see Agarwal *et al.*, 2003, Lemma 2.5). *A sequence of functions which is both uniformly bounded and equicontinuous in $\mathbb{T}$ contains a uniformly convergent subsequence.*

Theorem A.2 (Arzelà–Ascoli theorem; see Zhu *et al.*, 2007, Lemma 4). *A subset of $\mathcal{C}(\mathbb{T}, \mathbb{R})$ which is both equicontinuous and bounded is relatively compact. Here $\mathcal{C}(\mathbb{T}, \mathbb{R})$ denotes the class of real-valued continuous function defined on $\mathbb{T}$.*

Lemma A.1 (Urysohn lemma; see Granas and Dugungji, 2003, B.1 in Appendix). *Let X be a normal Hausdorff space, and let A and B be two disjoint nonempty closed subsets of X. Then there is a continuous map $\lambda \colon X \to [0, 1]$ such that $\lambda(a) = 1$ for $a \in A$ and $\lambda(b) = 0$ for $b \in B$.*

Theorem A.3 (Tychonoff fixed point theorem; see Granas and Dugungji, 2003, Theorem 1.10 in §7.1). *Let C be a nonempty compact convex set in a locally convex linear topological space X. Then every continuous map $F \colon C \to C$ has a fixed point.*

Lemma A.2 (see Hong and Peng, 2016, Lemma 4.12). *Let A be an $n \times n$-matrix-valued function in $\mathcal{R}_n$. If the homogeneous set* *dynamic equation*

$$\Delta_g U(t) = A(t)U(t), \quad t \in \mathbb{T},$$

admits an exponential dichotomy, then it has only one bounded solution $U(t) = 0$.

Proposition A.1 (see Lakshmikantham *et al.*, 2002, Proposition 1.3.4). *A nonempty subset A of the metric space (K^n, D) or (K_c^n, D), is compact if and only if it is closed and uniformly bounded.*

Lemma A.3 (see Coppel, 1978, Lemma 1 in Lecture 8). *Let A be a matrix-valued continuous function on $\mathbb{R}$. Let $X(t)$ be the fundamental matrix for*

$$x' = A(t)x, \quad t \in \mathbb{R}^+, \tag{A.1}$$

such that $X(0) = \mathcal{I}$. Suppose (A.1) has an exponential dichotomy

$$|X(t)PX^{-1}(s)| \le ke^{-\alpha(t-s)} \quad \text{if } t \ge s \tag{A.2}$$

$$|X(t)(\mathcal{I} - P)X^{-1}(s)| \le ke^{-\alpha(s-t)} \quad \text{if } s \ge t \tag{A.3}$$

for some positive constants k and α and a projection P, on $\mathbb{R}^+$. If, for some sequence $h_n \to \infty$, $A(t + h_n) \to B(t)$ uniformly on compact subintervals of $\mathbb{R}$, then $X(h_n)PX^{-1}(h_n) \to Q$ and the dynamic equation

$$y' = B(t)y, \quad t \in \mathbb{R},$$

has an exponential dichotomy on $\mathbb{R}$ with projection Q and the same constants k and α.

Lemma A.4 (see Li and Wang, 2011, Lemma 4.2). *Let S be an arbitrary compact subset of an open set in $\mathbb{E}^n$ ($\mathbb{R}^n$ or $\mathbb{C}^n$). If $f(t, x) \in \mathcal{C}(\mathbb{T} \times S, \mathbb{E}^n)$ is bounded on $\mathbb{T} \times S$, and the dynamic equation*

$$x^\Delta = f(t, x) \tag{A.4}$$

has a bounded solution ϕ such that $\{\phi(t) : t \in \mathbb{T}\} \subset S$ and $\mathbf{0} \in S$, then (A.4) must have a minimum norm solution.

Lemma A.5 (see Li and Wang, 2011, Lemma 4.3). *If $f(t, x)$ is almost periodic in t uniformly for $x \in \mathbb{E}^n$, $S = \overline{\{\phi(t): t \geq t_0\}}$ and (A.4) has a bounded solution ϕ on $[t_0, \infty)_{\mathbb{T}}$, then dynamic equation (A.4) has a bounded solution ψ on $\mathbb{T}$ and $\{\psi(t): t \in \mathbb{T}\} \subset S$.*

Lemma A.6 (see Li and Wang, 2011, Lemma 4.6). *Let $f(t, x) \in \mathcal{C}(\mathbb{T} \times \mathbb{E}^n, \mathbb{E}^n)$ be almost periodic in t uniformly for $x \in \mathbb{E}^n$ and, for every $G(t, x) \in H(f)$, the equation in the hull of (A.4),*

$$x^{\Delta} = G(t, x)$$

has a unique minimum norm solution. Then this minimum norm solution is almost periodic on $\mathbb{T}$.

Theorem A.4 (see Rhoades, 2001, Theorem 1). *Let E be a complete metric space with the distance $\hat{d}$ and $f: E \to E$ be a weakly contractive map, that is, there exists a continuous nondecreasing function $\psi: \mathbb{R}^+ \to \mathbb{R}^+$ with $\psi(0) = 0$, $\psi(t) > 0$ for $t > 0$, and*

$$\lim_{t \to \infty} \psi(t) = \infty$$

such that

$$\hat{d}(f(y), f(x)) \leq \hat{d}(x, y) - \psi(\hat{d}(x, y)).$$

Then, f has a unique fixed point in E.

Below we state the Arzelà–Ascoli theorem for set-valued mappings.

Theorem A.5 (see Lakshmikantham *et al.*, 2006, Theorem 2.4.1). *If $\{U_n\}$ is a sequence of equicontinuous and equibounded set-valued functions defined on an interval J, then we can extract a subsequence that converge uniformly to a continuous set-valued function U on J.*

Theorem A.6 (see Stefanini and Bede, 2009, Theorem 17). *Let $f: [a, b] \to \mathbb{I}$ be such that $f(x) = [f^-(x), f^+(x)]$. The function f is generalized Hukuhara differentiable if and only if f^- and f^+ are differentiable real-valued functions. Furthermore,*

$$f'(x) = \left[\min\left\{(f^-)'(x), (f^+)'(x)\right\}, \max\left\{(f^-)'(x), (f^+)'(x)\right\}\right],$$
$$x \in [a, b].$$

Theorem A.7 (Comparison result; see Kaymakcalan, 1993, Theorem 5.2). *Let $B = \{u \in \mathbb{R}: |u - u_0| \leq b,\ b > 0\}$ and $R_0 = [t_0, t_0 + a] \times B$ and, let $g \in \mathcal{C}_{\mathrm{rd}}(R_0, \mathbb{R})$ be such that $g(t, u)\mu(t)$ be nondecreasing in u for each $t \in [t_0, t_0+a]_{\mathbb{T}}$. Suppose $m: [t_0, t_0+a)_{\mathbb{T}} \to \mathbb{R}$ be delta differentiable such that*

$$m^{\Delta}(t) \leq g(t, m(t)) \quad \text{for } t \in [t_0, t_0 + a)_{\mathbb{T}}.$$

Then, $m(t_0) \leq u_0$ implies that

$$m(t) \leq r(t) \quad \text{for } t \in [t_0, t_0 + a)_{\mathbb{T}},$$

where $r(t)$ is the maximal solution of

$$u^{\Delta}(t) = g(t, u) \quad \text{for } u(t_0) = u_0$$

existing on $[t_0, t_0 + a)_{\mathbb{T}}$.

Theorem A.8 (see Kaymakcalan, 1993, Theorem 6.2). *Assume that*

(i) $f \in \mathcal{C}_{\mathrm{rd}}(\mathbb{T} \times \mathbb{R}^n, \mathbb{R}^n)$, $g \in \mathcal{C}_{\mathrm{rd}}(\mathbb{T} \times \mathbb{R}, \mathbb{R})$, $f(t, 0) = 0$, $g(t, 0) = 0$, *and for* $(t, x) \in \mathbb{T} \times \mathbb{R}^n$

$$[x, f(t, x)]_+ \equiv \lim_{h \to 0} \frac{1}{h}\left[|x + hf(t, x)| - |x|\right] \leq g(t, |x|);$$

(ii) $g(t, u)\mu(t)$ *is nondecreasing in u for each $t \in \mathbb{T}$.*

Then the stability properties of the trivial solution of

$$u^{\Delta} = g(t, u), \quad u(t_0) = u_0 \geq 0, \quad t \in \mathbb{T}$$

implies the corresponding stability properties of the trivial solution of

$$x^{\Delta} = f(t, x), \quad x(t_0) = x_0 \quad \text{with } \|x_0\| \leq u_0.$$

Theorem A.9 (see Li and Hong, 2011, Theorem 4.2). *Assume that $V(t, U(t)) \in \mathcal{C}^1_{\mathrm{rd}}(\mathbb{T} \times K^1_c, \mathbb{R}^+)$ for $U \in \mathcal{C}_{\mathrm{rd}}(\mathbb{T}, K^1_c)$ satisfies the following conditions.*

(i) *There exist positive functions λ_1, λ_2 and positive constants p, q such that*

$$\lambda_1(t)\|U(t)\|^p \leq V(t, U(t)) \leq \lambda_2(t)\|U(t)\|^q.$$

(ii) *There exist a positive function λ_3, nonnegative constant L, positive constant r and constant α with*

$$\alpha > M := \inf_{t \geq 0} \frac{\lambda_3(t)}{[\lambda_2(t)]^{r/q}} > 0$$

such that

$$V^\Delta(t, U(t)) \leq \frac{-\lambda_3(t)\|U(t)\|^r - L(M \ominus \alpha)(t)e_{\ominus \alpha}(t, 0)}{1 + M\mu(t)}.$$

(iii) $$V(t, U(t)) - (V(t, U(t)))^{r/q} \leq 0,$$

where q and r are constants given as in (i) and (ii), respectively.

Then the trivial solution of SDIVP

$$\Delta_H U = F(t, U), \quad U(t_0) = U_0 \in K_c^1,$$

is exponentially stable on $\mathbb{T}$.

Lemma A.7 (see Hong *et al.*, 2014, Lemma 12). *Let $A \in \mathcal{R}$, $F \in \mathcal{C}_{\mathrm{rd}}(\mathbb{T}, K_c^n)$, $t_0 \in \mathbb{T}$, and $U_0 \in K_c^n$. Then the SDIVP*

$$\Delta_g U(t) = A(t)U(t) + F(t) \quad \text{for } t \in \mathbb{T},$$
$$U(t_0) = U_0,$$

has a unique solution $\mathscr{U} : \mathbb{T} \to K_c^n$ given by

$$\mathscr{U}(t, t_0, U_0) = e_A(t, t_0)U_0 + \int_{t_0}^t e_A(t, \sigma(s))F(s)\Delta s.$$

Lemma A.8 (see Hong *et al.*, 2014, Lemma 13). *Let $A \in \mathcal{R}$, $F \in \mathcal{C}_{\mathrm{rd}}(\mathbb{T}, K_c^n)$, $t_0 \in \mathbb{T}$, and $U_0 \in K_c^n$. Then the SDIVP*

$$\Delta_g U(t) = -A^*(t)U(\sigma(t)) + F(t) \quad \text{for } t \in \mathbb{T},$$
$$U(t_0) = U_0,$$

has a unique solution $\mathscr{U}^* : \mathbb{T} \to K_c^n$ given by

$$\mathscr{U}^*(t, t_0, U_0) = e_{\ominus A^*}(t, t_0)U_0 + \int_{t_0}^t e_{\ominus A}(t, s)F(s)\Delta s.$$

Proposition A.2 (see Hong, 2009, Proposition 3.11). *Assume that $t_0, T \in \mathbb{T}$ and $F, G \colon [t_0, T]_\mathbb{T} \to K_c^1$ are Δ_H-integrable and have rd-continuous selectors. Then, we have*

(i) $\int_{t_0}^T [F(s) + G(s)]\Delta s = \int_{t_0}^T F(s)\Delta s + \int_{t_0}^T G(s)\Delta s.$

(ii) $\int_{t_0}^t \lambda F(s)\Delta s = \lambda \int_{t_0}^t F(s)\Delta s, \quad \lambda \in \mathbb{R}^+, \quad t \in [t_0, T]_\mathbb{T}.$

(iii) $\int_{t_0}^T F(s)\Delta s = \int_{t_0}^t F(s)\Delta s + \int_t^T F(s)\Delta s, \quad t \in [t_0, T]_\mathbb{T},$

      $t_0 \le t \le T.$

(iv) $\int_{t_0}^{t_0} F(s)\Delta s = \{0\}.$

(v) *If* $f \in \mathcal{S}_F([t_0, T]_\mathbb{T})$ *implies that* $f \in \mathcal{C}_{\mathrm{rd}}([t_0, T]_\mathbb{T})$, *then* $\|F(\cdot)\| \colon [t_0, T]_\mathbb{T} \to \mathbb{R}^+$ *is delta integrable and*

$$\left\| \int_{t_0}^T F(s)\Delta s \right\| \le \int_{t_0}^T \|F(s)\|\Delta s.$$

(vi) *If* $f \in \mathcal{S}_F([t_0, T]_\mathbb{T})$ *and* $g \in \mathcal{S}_G([t_0, T]_\mathbb{T})$ *implies that* $f \in \mathcal{C}_{\mathrm{rd}}([t_0, T]_\mathbb{T})$ *and* $g \in \mathcal{C}_{\mathrm{rd}}([t_0, T]_\mathbb{T})$, *respectively, then* $D[F(\cdot), G(\cdot)] \colon [t_0, T]_\mathbb{T} \to \mathbb{R}^+$ *is delta integrable and*

$$D\left[\int_{t_0}^T F(s)\Delta s, \int_{t_0}^T G(s)\Delta s \right] \le \int_{t_0}^T D[F(s), G(s)]\Delta s.$$

Bibliography

R. P. Agarwal, M. Bohner, and P. Řehák, Half-linear dynamic equations, *Nonlinear Analysis and Applications* (*to V. Lakshmikantham on his 80th birthday*), **1–2** (2003), 1–57.

R. P. Agarwal, B. Hazarika, and S. Tikare (eds.) *Dynamic Equations on Time Scales and Applications*. CRC Press, Boca Raton FL, 2024.

B. Ahmad and S. Sivasundaram, The monotone iterative technique for impulsive hybrid set valued integro-differential equations, *Nonlinear Analysis*, **65**(12) (2006), 2260–2276.

F. M. Atici, G. Sh. Guseinov, and B. Kaymakçalan, On Lyapunov inequality in stability theory for Hill's equation on time scales, *Journal of Inequalities and Applications*, **5**(6) (2000), 603–620.

B. Bede and S. G. Gal, Almost periodic fuzzy-number-valued functions, *Fuzzy Sets and Systems*, **147**(3) (2004), 385–403.

B. Bede and S. G. Gal, Generalizations of the differentiability of fuzzy-number-valued functions with applications to fuzzy differential equations, *Fuzzy Sets and Systems*, **151**(3) (2005), 581–599.

B. Bede, I. J. Rudas, and A. L. Bencsik, First order linear fuzzy differential equations under generalized differentiability, *Information Sciences*, **177**(7) (2007), 1648–1662.

T. G. Bhaskar and M. Shaw, Stability results for set difference equations *Dynamic Systems and Applications*, **13**(3&4) (2004), 479–485.

T. G. Bhaskar and J. Vasundhara Devi, Stability criteria for set differential equations, *Mathematical and Computer Modelling*, **41** (2005), 1371–1378.

L. Bi, M. Bohner, and M. Fan, Periodic solutions of functional dynamic equations with infinite delay, *Nonlinear Analysis TMA*, **68**(5) (2008), 1226–1245.

S. Bochner, A new approach to almost periodicity, in *Proceedings of the National Academy of Sciences of the United States of America*, **48**(12) (1962), 2039–2043.

M. Bohner and A. Peterson, *Dynamic Equations on Time Scales, An Introduction with Applications*, Birkhäuser, Boston, 2001.

M. Bohner and A. Peterson (eds.), *Advances in Dynamic Equations on Time Scales*, Birkhäuser, Boston, 2003.

A. J. Brandao Lopes Pinto, F. S. De Blasi, and F. Iervolino, Uniqueness and existence theorems for differential equations with compact convex valued solutions, *Boiletino dell Unione Matematica Italiana*, **3** (1970), 47–54.

Y. Chalco-Cano and H. Román-Flores, On new solutions of fuzzy differential equations, *Chaos, Solitons and Fractals*, **38**(1) (2008), 112–119.

Y. Chalco-Cano and H. Román-Flores, Comparison between some approaches to solve fuzzy differential equations, *Fuzzy Sets and Systems*, **160**(11) (2009), 1517–1527.

Y. Chalco-Cano, H. Román-Flores, and M. D. Jiménez-Gamero, Generalized derivative and π-derivative for set-valued functions, *Information Sciences*, **181**(11) (2011), 2177–2188.

D. Cheban and C. Mammana, Invariant manifolds, global attractors and almost periodic solutions of nonautonomous difference equations, *Nonlinear Analysis TMA*, **56**(4) (2004), 465–484.

W. A. Coppel, *Dichotomies in Stability Theory*, Lecture Notes in Mathematics, 629, Springer Berlin, Heidelberg, 1978.

C. Corduneanu, *Almost Periodic Functions*, second edn., Chelsea, New York, 1989.

P. Diamond and P. Kloeden, *Metric Spaces of Fuzzy Sets*, World Scientific, Singapore, 1994.

S. G. Georgiev, *Fuzzy Dynamic Equations, Dynamic Inclusions, and Optimal Control Problems on Time Scales*, Springer Cham, 2021.

S. G. Georgiev, *Dynamic Calculus and Equations on Time Scales*, De Gruyter, Berlin, Boston, MA, 2023.

A. Granas and J. Dugundji, *Fixed Point Theory*, Springer Monographs in Mathematics, Springer-Verlag, New York, 2003.

S. Hilger, Analysis on measure chains — a unified approach to continuous and discrete calculus, *Results in Mathematics*, **18** (1990), 18–56.

S. H. Hong, Differentiability of multivalued functions on time scales and applications to multivalued dynamic equations, *Nonlinear Analysis TMA*, **71**(9) (2009), 3622–3637.

S. H. Hong, Stability criteria for set dynamic equations on time scales, *Computers and Mathematics with Applications*, **59**(11) (2010), 3444–3457.

S. H. Hong, Phase spaces and periodic solutions of set functional dynamic equations with infinite delay, *Nonlinear Analysis TMA*, **74**(9) (2011), 2966–2984.

S. H. Hong, J. Gao, and Y. Peng, Solvability and stability of impulsive set dynamic equations on time scales, *Abstract and Applied Analysis*, ID 610365 (2014), 1–19.

S. H. Hong and Y. Peng, Almost periodicity of set-valued functions and set dynamic equations on time scales, *Information Sciences*, **330** (2016), 157–174.

S. H. Hong, X. Cao, J. Chen, H. Hou, and X. Luo, General forms of solutions for linear impulsive fuzzy dynamic equations on time scales, *Discrete Dynamics in Nature and Society*, Article ID 4894921, **2020** (2020), 1–19.

M. Hukuhara, Sur l'aplication semicontinuone dont la valeur est un compact convexe, *Funkcial. Ekvak.* **10** (1967), 43–68.

K. Jia, S. H. Hong, X. Cao, and J. Yue, Exponential stability for a class of set dynamic equations on time scales, *Journal of Inequalities and Applications*, **2022**(135) (2022), 1–18.

B. Kaymakcalan, Existence and comparison results for dynamic systems on a time scale, *Journal of Mathematical Analysis and Applications*, **172**(1) (1993), 243–255.

B. Kaymakcalan, Stability analysis in terms of two measures for dynamic systems on time scales, *Journal of Applied Mathematics and Stochastic Analysis*, **6**(4) (1993), 325–344.

A. Khastan, J. J. Nieto, and R. R. López, Variation of constant formula for first order fuzzy differential equations, *Fuzzy Sets and Systems*, **177**(1) (2011), 20–33.

V. Lakshmikantham, S. Leela, and A. S. Vatsala, Set valued hybrid differential equations and stability in terms of two measures, *Nonlinear Analysis: Hybrid Systems*, **2**(2) (2002), 169–188.

V. Lakshmikantham, S. Leela, and J. Vasundhara Devi, Stability theory for set differential equations, *Dynamics of Continuous, Discrete and Impulsive Systems Series A: Mathematical Analysis*, **11** (2004), 181–189.

V. Lakshmikantham, T. Gnana Bhaskar, and J. V. Devi, *Theory of Set Differential Equations in Metric Spaces*, Cambridge Scientific Publishers, 2006.

L. Li and S. H. Hong, Exponential stability for set dynamic equations on time scales, *Journal of Computational and Applied Mathematics*, **235** (2011), 4916–4924.

J. Li, A. Zhao, and J. Yan, The Cauchy problem of fuzzy differential equations under generalized differentiability, *Fuzzy Sets and Systems*, **200** (2012), 1–24.

Y. Li and C. Wang, Uniformly almost periodic functions and almost periodic solutions to dynamic equations on time scales, *Abstract and Applied Analysis*, Article ID 341520 (2011), 1–22.

Y. Li and C. Wang, Almost periodic solutions of shunting inhibitory cellular neural networks on time scales, *Communications in Nonlinear Science and Numerical Simulation*, **17**(8) (2012), 3258–3266.

A. A. Martynyuk, *Qualitative Analysis of Set-valued Differential Equations*, Springer Nature, Switzerland AG, 2019.

A. A. Martynyuk, I. M. Stamova, and A. Yu, Martynyuk-Chernienko, Matrix Lyapunov functions method for sets of dynamic equations on time scales, *Nonlinear Analysis: Hybrid Systems*, **34** (2019) 166–178.

B. E. Rhoades, Some theorems on weakly contractive maps, *Nonlinear Analysis TMA*, **47**(4) (2001), 2683–2693.

L. Stefanini, A generalization of Hukuhara difference for interval and fuzzy arithmetic, in *Soft Methods for Handling Variability and Imprecision*, D. Dubois, M. A. Lubiano, H. Prade, M. A. Gil, P. Grzegorzewski, and O. Hryniewicz, (eds.), vol. 48 of Series on Advances in Soft Computing, Springer, New York, NY, USA, 2008, http://econpapers. repec.org/RAS/pst233.htm.

L. Stefanini and B. Bede, Generalization of Hukuhara differentiability of interval-valued functions and interval differential equations, *Nonlinear Analysis TMA*, **71**(3–4) (2009), 1311–1328.

S. Tikare and C. C. Tisdell, Nonlinear dynamic equations on time scales with impulses and nonlocal initial conditions, *Journal of Classical Analysis*, **16**(2) (2020), 125–140.

C. Vasavi, G. Suresh Kumar, and M. S. N. Murty, Generalized differentiability and integrability for fuzzy set-valued functions on time scales, *Soft Computing*, **20**(3) (2016), 1093–1104.

J. Zhang, M. Fan, and H. Zhu, Existence and roughness of exponential dichotomies of linear dynamic equations on time scales, *Computers and Mathematics with Applications*, **59**(8) (2010), 2658–2675.

Z.-Q. Zhu and Q.-R. Wang, Existence of nonoscillatory solutions to neutral dynamic equations on time scales, *Journal of Mathematical Analysis and Applications*, **335**(2) (2007), 751–762.

Index

www.ingramcontent.com/pod-product-compliance
Lightning Source LLC
Chambersburg PA
CBHW050637190326
41458CB00008B/2314